ruins

THE CONSERVATION AND REPAIR OF MASONRY RUINS

Comhshaol, Oidhreacht agus Rialtas Áitiúil
Environment, Heritage and Local Government

DUBLIN
PUBLISHED BY THE STATIONERY OFFICE
To be purchased directly from:
Government Publications Sales Office
Sun Alliance House
Molesworth Street
Dublin 2

or by mail order from:
Government Publications
Postal Trade Section
Unit 20 Lakeside Retail Park
Claremorris
Co. Mayo

Tel: 01 - 6476834/37 or 1890 213434; Fax: 01 - 6476843 or 094 - 9378964
or through any bookseller

© Government of Ireland 2010
ISBN 978-1-4064-2445-4

All or part of this publication may be reproduced without further permission provided the source is
acknowledged. The Department of the Environment, Heritage and Local Government and the authors
accept no liability for any loss or damage resulting from reliance on the advice contained in this booklet

Text by: Margaret Quinlan with Mary Hanna and David Kelly
Contributors: John Feehan, Willie McErlean, Dave Pollock, Aidan Smith
Editorial team: Aighleann O'Shaughnessy, Margaret Keane, Jacqui Donnelly
All images are by the authors or DoEHLG, unless otherwise stated

Series Editor: Jacqui Donnelly
Copy Editor: Anna Kealy
Design: Bennis Design

Cover image: Abbey and Hospital of St Mary d'Urso, Drogheda, County Louth

Contents

INTRODUCTION	**5**
1. TRADITIONAL MASONRY CONSTRUCTION IN IRELAND	**8**
Prehistoric stone structures	8
Mediaeval masonry buildings	8
Post-mediaeval masonry buildings	10
Ruins found today	12
Materials	13
Types of masonry construction	16
Features of masonry walls	19
2. APPROACH TO MAINTENANCE AND REPAIR	**27**
Recording and assessment by the owner or custodian	27
Checklist for owners or custodians	29
Getting the right advice	29
What to expect from a professional advisor	30
Level of recording	30
Preventive maintenance	31
3. COMMON DEFECTS FOUND IN RUINED STRUCTURES	**34**
Damage from vegetation	34
Foundation failures	35
Defects in walls	35
4. PLANNING FOR REPAIR	**44**
Principal materials used in repair	44
Temporary works	47
Providing access	48
Loose carved stone and other artefacts	49
Dealing with vegetation	49
5. REPAIR TECHNIQUES	**51**
Underpinning foundations	51
Wall cappings	51
Stitching	52
Rebuilding collapsed masonry	52
Partial dismantling and rebuilding	52
Repairs to leaning walls	53
Repairs to failing arches	54

Repairs to failing lintels	54
Replacement stones	54
Repointing	55
Grouting rubble masonry	55
Protection of carved stone and other decorative features	56
Shelter coats	56
Surviving plaster, render or slate-hanging	56
Surviving timber or evidence of timber	56
Surviving floors or paving	57
Dealing with emergencies such as collapse	57
Distinguishing between old and new in repair work	57
The use of consolidants	58
Stone cleaning	58
Lichen growth on stone	58

6. ARCHAEOLOGY — 59

Archaeological testing and monitoring requirements	59
Sub-surface archaeology	59
Site works	59
Removal of tree roots	60
Collapsed masonry	60

7. OTHER IMPORTANT ISSUES — 64

Setting	64
Floodlighting	64
Lightning protection	65
Ecological and wildlife issues	65
Farming practices	66
Structures accessible to the public	66
Burials	66
Reuse of ruins	67

GLOSSARY — 70

Introduction

It will have been seen that this is a country of ruins. Lordly or humble, military or domestic, standing up with furious gauntness…or shelving weakly into the soil, ruins feature the landscape – uplands or river valleys – and make a ghostly extra quarter to towns. They give clearings in woods, reaches of mountain or sudden turns of a road a meaning and pre-inhabited air. Ivy grapples them; trees grow inside their doors; enduring ruins, where they emerge from ivy, are the limestone white-grey and look like rocks.

From *Bowen's Court* by Elizabeth Bowen, 1942

Ruins are the surviving remnants of structures whose function came to an end some time in the past. In many cases these remnants bear witness to the manner of that end: for some it was accidental; for others it was violent, through deliberate destruction in one of the many episodes of warfare that punctuate the story of Ireland; or through a political act such as the dissolution of the monasteries.

For others the evidence is different: the unroofed house, the clachán or the abandoned shell of a mill tell of changed economic circumstances that made their function redundant or upkeep impossible. Often, the later grander building nearby tells of an increase in prosperity rather than decline. The adaptations of the buildings of the past are themselves records of change and illustrate the scale of transition visible in the landscape of this country. Each survival is a document that embodies a social, cultural and economic history.

Interest in the ruins of the past has deep roots. The Enlightenment of the eighteenth century laid the foundation for the development of antiquarian interest in the built relics of the past, leading in time to the formal study of archaeology and architecture. This interest was overlaid with appreciation of the picturesque. In the eighteenth and nineteenth centuries, antiquarian illustrators and scholars - Beranger, Grose, Petrie and Du Noyer among others - documented the views, buildings and antiquities of the Irish countryside through drawings, paintings and prints that evoked a magical and sometimes desolate land. Ivy-clad ruined castles, monasteries and forsaken churches convey a picture of melancholic beauty that accorded with the romantic spirit of the times.

Interest in the study of ancient ruins grew throughout the second half of the eighteenth century and into the nineteenth century. Many artists and antiquarians published collections of topographical images which are now invaluable sources of reference. This drawing of Jerpoint Abbey, County Kilkenny by Edward Cheney is dated 1837 (Image courtesy of the Office of Public Works)

Ruins became a prized feature of the designed Romantic landscape to the point where, if one did not have a convenient authentic ruin, a sham could be constructed as an 'eye catcher' or focal point in a vista. Towers and temples modelled on icons of classical architecture and built as elements in the picturesque landscape are nowadays often themselves ruins. Mediaeval buildings, in particular, acquired an almost moral status, representing an imagined ideal past and offering a model of hope for the future. This status was also expressed in the nineteenth century by the adoption of mediaeval styles for new church buildings, as well as in the restoration and rebuilding of existing churches. At times, the works of the past became a means of underwriting cultural identity for different religious denominations.

But the ruin has more to offer than lessons in history, cultural identity or aesthetics. The broken walls of an abbey or former mill have much to tell, not only about the ambitions of its patron, but also about historic skills and building methods. In our time, these remains have reminded us of the discarded wisdom of the past, such as the importance of setting in the landscape and the use of lime mortars and renders.

Our purpose here is to help owners to retain these ruins as documents, as evidence, as object lessons, as things of beauty and as inspirations, and to do so without changing their essential character. They retain intrinsic archaeological, architectural and historic importance. Many have become associated with the surrounding landscape and are beautiful in themselves.

Some ruined structures, because of their national importance, are protected by being in the ownership or guardianship of the State; these are mostly structures dating from before 1700 AD. Bodies such as local authorities and religious authorities may also be the owners of many ruined structures, particularly mediaeval or later churches, where ownership includes the building and its surrounding graveyard. Some community groups have been formed to take responsibility for a neglected historic feature in their area which has local significance.

Lynally Glebe, County Offaly. Many rural graveyards in local authority ownership or control contain the ruins of early mediaeval churches which are included in the Record of Monuments and Places

There are thousands of ruined structures in the ownership of individuals. These can range from tower houses, grain stores and mills to limekilns and vernacular farm buildings. Depending on the level of protection afforded to these ruins under planning and development and/or national monuments legislation, their owners and custodians have various responsibilities for repairing and maintaining them.

Ruined tower houses can be found isolated in the landscape or nestled in the grounds of the houses that replaced them

This guide has been compiled to assist owners and custodians of ruined structures to become aware of their condition, potential dangers and vulnerability to various types of damage and to take appropriate steps to secure the ruin into the future. In approaching the conservation of a ruined structure, a key decision will be the level of professional expertise required, whether that of an architect, engineer, archaeologist, skilled workforce or a combination of these. The decision will depend on the age and importance of the structure, its condition and the level of work required. This publication does not offer comprehensive technical advice. Nor does it provide advice on the health and safety issues which arise in the repair, maintenance and conservation of these structures. Each ruined structure is unique, and there are no standard solutions for its repair and conservation. Each structure merits individual consideration; and its particular history, context, development and previous repairs should be used to guide the correct conservation and repair methodologies.

Many vernacular buildings survive as ruins. This well-built stone building in Mannin, County Galway, is dated 1786. Such modest buildings are a vulnerable part of the landscape, having outlived their purpose, and are often without statutory protection

1. Traditional Masonry Construction in Ireland

Prehistoric stone structures

Stone has been used as a building material in Ireland for almost six thousand years. The earliest surviving examples of stone construction date from the Neolithic period. While the majority of structures erected in prehistoric times were made of timber and earth, the only ones that survive in anything resembling their original form are those that were built in stone. Stone was used for the construction of tombs and sacred sites, of which there are magnificent remains at monuments such as Newgrange, built about 5,000 years ago. Also dating from the Neolithic period, the most extensive remains of field boundary walls, tombs and enclosures yet found are at the Céide Fields in County Mayo. The Late Bronze Age saw the construction of the great stone forts on the western seaboard, such as Dún Aonghusa on the Aran Islands, which was inhabited around 1,300 BC. These masonry structures were of drystone construction incorporating such architectural features as enclosing walls, internal terraces accessed by stone steps, passages and chambers within the wall widths and defensive chevaux de frise outside the ramparts.

The early mediaeval monastery of Skellig Michael, off the coast of County Kerry included a series of beehive-shaped cells of drystone construction built to withstand the harsh climate at the edge of the Atlantic ocean. Circular in shape from the outside, the internal spaces of the cells are in fact square or oblong

Mediaeval masonry buildings

With the introduction of Christianity in the fifth century, Ireland was opened up to new influences from continental Europe, including in architecture and building technology. Most early mediaeval buildings in this country were constructed of perishable materials such as timber, wattle or clay. None of these structures survive upstanding. Drystone construction, that is without the use of mortar, was used at a slightly later date to build the walls and sometimes even the roofs of buildings especially in County Kerry. The principles of corbelling were well-known from earlier times, corbelled stone having been used to roof the great passage tombs of the Neolithic period. The creation of a corbelled vault involved building up a succession of stone courses, each of which stepped inwards from the one below it until they met at the centre. Corbelling relied on the careful selection, partial working and setting of stones, sloping to the exterior, to divert rainwater away from the inside of the building.

Even in those parts of the island where there was an abundance of timber, there was a desire to build more permanent buildings for worship. The use of lime-based mortars to bind masonry units together had been widespread throughout the Roman Empire but they were not used in Ireland until a relatively late date, possibly as late as the eighth century. The use of mortar to bind stones together in a masonry wall allowed for stronger, more stable structures to be built using less material.

Early masonry churches built of mortared masonry were simple, rectangular structures, varying in size. They were built of local stone roughly dressed, if at all, but with more carefully worked stones used for the quoins (or corner stones) and around the openings.

The stark simplicity of the early mediaeval oratory of Temple Benan on Inishmore shows a sophisticated use of local stone including large thin blocks of limestone laid on edge

Sherkin Island, County Cork. The high mediaeval period saw masonry techniques and skills evolve to allow the construction of larger and more complex stone buildings than had been seen before in Ireland. Many of the religious buildings of this period were abandoned following the Dissolution of the Monasteries in 1537

From the tenth to the twelfth century, the techniques and understanding of stone construction improved. The introduction of the arch into Irish masonry construction marked another advance in building technology and had a profound influence on architectural style. The Romanesque style, based on the round arch, is mainly associated in Britain with the buildings of the Anglo-Normans. However, in Ireland the style was favoured by the Gaelic culture, and Irish Romanesque buildings generally predate the Anglo-Norman invasion of 1169, in some cases by several decades. The period from the tenth to the twelfth century also saw the construction of round towers. Later used in the Celtic Revival period to symbolise Irish nationhood, these slender and sophisticated masonry structures are tapered in profile rising to heights of 25m or more.

Towards the end of the twelfth century, the Romanesque style began to give way to the Gothic, particularly in the design of religious buildings. Characterised by pointed arches, the Gothic style pushed masonry technology to its limits. With the aid of new building elements such as flying buttresses and pinnacles, stone buildings were constructed that were higher and spanned greater distances than ever before.

Between the religious and the secular, the thirteenth century was a period of great building activity; early timber fortifications were replaced by stone castles and defences. Trim Castle and Carrickfergus Castle are outstanding examples of the great military castles constructed in this period, while many towns retain remnants of extensive stone defensive walls in varying states of preservation.

The town walls of Athenry, County Galway are among the most intact and impressive circuits of town defences in the country

The principal era of stone-castle building was between the end of the twelfth and the middle of the fourteenth century. Most later castles built in Ireland took the form of tower houses, which were the fortified residences of the land-owning classes, both Irish and Anglo-Norman. It is estimated that some seven thousand tower houses were built between the fourteenth and seventeenth centuries. While many were remodelled and incorporated into later buildings and farms, most tower houses were later abandoned and now stand as ruins in the landscape.

Fortified houses were not confined to rural locations; this urban tower house was one of several built within the mediaeval walled town of Fethard, County Tipperary

While the iconic image of a tower house is a ruin standing alone in the landscape, many were originally surrounded by bawns and other buildings. A surprising number survive integrated into later houses or, as in this example, with houses added on to them. The outline of the roof of an earlier attached house can be seen on the wall of the tower house

Post-mediaeval masonry buildings

In the late sixteenth and early seventeenth centuries, many of the dwellings of the gentry no longer required defensive features and a less fortified style of architecture emerged such as can be seen at Rathfarnham Castle, County Dublin and Portumna Castle, County Galway. Masonry walls decreased in thickness as they no longer needed to be constructed to withstand artillery bombardment; instead the need to keep driving rain from the interior of the building became the determining factor. By the end of the seventeenth century, defensive features such as crenellations, machicolations and the like were no longer required, although gun loops are occasionally found incorporated into buildings as late as the eighteenth century.

At Loughmoe Court, County Tipperary, the original fifteenth century tower house (seen on the left) was extended in the first decades of the seventeenth century with the addition of a semi-fortified Jacobean mansion

The eighteenth century saw the development of buildings made of fine ashlar stone or of handmade brick, with classical detailing following European styles of architecture but often with distinctly Irish characteristics. External walls were often built of masonry rubble which was then rendered, sometimes including incised lines to give the appearance of ashlar.

The impressive remains of Tyrone House, County Galway, built in 1779 and abandoned in 1905. Tyrone House and its occupants were the inspiration for the Somerville & Ross novel 'The Big House of Inver'

The skills and techniques of masonry building became more advanced throughout the nineteenth century. There was a revival of interest in mediaeval architecture and building techniques, particularly amongst architects, and a desire to return to mediaeval building practices. Architects and antiquarians began to study in detail the ancient ruins that survived throughout the countryside to learn from their construction methods and architectural styles. The Gothic Revival style grew out of this interest in ancient buildings and its influence can be seen in a variety of building types including churches, banks and country houses. Architects of the period also experimented with designs in other historical styles such as Tudor, Jacobean and Queen Anne Revival.

The Board of First Fruits of the Church of Ireland built a large number of rural churches in a distinctive Georgian-Gothic style during the first decades of the nineteenth century. While many of these churches are still in use today, others fell into disuse. Even as ruins, they continue to play a role as prominent local landmarks

As well as buildings of high architectural quality, vernacular masonry structures were plentiful throughout the countryside; simple farms and homes were built with local rubble stone finished in lime render, sometimes with distinctive regional variations. Industrial buildings, including wind and water mills, saw mills, brickworks and limekilns, were commonplace even in areas today not associated with industrialisation.

Ruins found today

Political unrest and changes in land tenure in the late-nineteenth and early-twentieth centuries led to the destruction and abandonment of many country houses, the ruins of which still stand. Other cultural and societal changes led to the un-roofing and even the complete or partial demolition of many rural churches. Evolving industrial practices meant that many buildings such as mills, breweries or farm buildings could no longer accommodate the purposes for which they were built and were abandoned in favour of new buildings.

Limekilns were once a common feature throughout the countryside, burning limestone to provide quicklime for use in building mortars or as a fertiliser

Ruins of all these building types remain throughout the country and are an important part of our national heritage. While they have survived, sometimes for hundreds of years, they will not survive indefinitely without periodic care and attention. Ruins are buildings that no longer have their original defences against the elements, such as roofs, windows and doors, while the loss of floors, parts of walls and other fabric may have altered or weakened their structural integrity.

TYPES OF RUINED STRUCTURES

The types of ruined structures discussed in this booklet mainly span the period from early mediaeval churches to the industrial buildings of the nineteenth century. The range of buildings is very wide and broadly includes the following:

AGRICULTURAL BUILDINGS

Barns, bee-boles, booley huts, gate piers, yard buildings

BURIAL STRUCTURES

Burial grounds, churchyards, funerary sites and monuments, graveyards, mausolea

CIVIC BUILDINGS

Courthouses, market houses, town walls, work-houses, schools and other educational buildings

DWELLINGS

Castles and tower houses, mansions and associated buildings such as coach houses, outbuildings and stables, vernacular dwellings

INDUSTRIAL AND COMMERCIAL BUILDINGS

Breweries, brickworks, bridges, canal buildings, chimneys, distilleries, glass houses, grain stores, industrial chimneys, lime kilns, maltings, mills, mine engine houses, pump houses, railway buildings, tanneries, windmills

LANDSCAPE FEATURES

Dovecotes, follies, gateways and gate lodges, gazebos, ice houses, obelisks, summer houses, walled gardens

MILITARY/DEFENCE AND FORTIFIED STRUCTURES

Barracks, batteries, forts, fortifications, gate houses, Martello towers, mural towers, town defences, watchtowers

RELIGIOUS BUILDINGS

Abbeys, almshouses, cathedrals, chapels, churches, convents, oratories, priories, religious houses

Materials

Traditional masonry construction consists of stone or brick with lime or clay mortar or, less commonly, drystone construction without mortar. Stone, being a heavy material, was not easy to transport in the past and even transport from nearby quarries by horse and cart was difficult. As a result, many historic stone structures reflect the geology of the immediate area. Small quarries and limekilns existed in every part of the country.

Where there was a requirement for carved stone, and a suitable stone was not locally available, it was transported from another source, generally by water. Imported limestone such as Dundry stone from Somerset in England and Caen stone from Normandy in France were used in parts of the country for mediaeval carved work. By the eighteenth century, stone such as Portland stone from Dorset was imported for use in high quality buildings, either by itself as a wall-facing material or for carved dressings used in combination with other building materials. Increased industrialisation in the nineteenth century meant that a variety of different types of imported stone was readily available throughout the country.

Dundry stone, a cream-coloured oolitic limestone, was popular with mediaeval builders because it could be easily carved. It was imported from Somerset for use in the decorative features of many castles and ecclesiastical buildings

STONE

Stone is classified into three main types, depending on how it was formed:

> Igneous
> Sedimentary
> Metamorphic

Igneous stone

Igneous rocks originate as magma. Where this magma cools slowly beneath the surface, a coarse-textured igneous rock, such as granite, forms. Where it reaches the surface as lava and then cools, a finer-grained igneous rock such as basalt is the result.

Granite was the chief igneous stone used for building. Generally found in mountainous areas, it was used either as rubble stone or cut stone. It is probably the most robust of all stones because of its hardness and durability. However, its hardness made it difficult to work. Due to its coarse composition, it was not suitable for fine carving and was used mostly in simple shapes.

Sedimentary stone

Sedimentary rocks are formed through the accumulation of mineral and organic particles deposited by water, ice or wind and usually laid down in layers. Sedimentary stones used for building purposes include limestones and sandstones. They are usually softer than granite.

Limestone was by far the most widely used stone in historic masonry construction in this country, being readily available and, for the most part, relatively easy to work. Many Irish towns and cities were built from limestone which was quarried locally. Limestone was also used in the making of lime for mortars and plasters. It remains a major source of dimensioned stone, aggregate and lime, and a constituent of cement.

Sandstone can be hard or soft, depending on the binding material. In the distinctive stone known as Old Red Sandstone, which occurs widely in Ireland, the binder is silica. Sandstones used for carving and decorative work were softer and easier to work. However, they are also more susceptible to decay from rain and wind.

Granite ashlar with vermiculation to the lower course

Limestone ashlar in a masonry wall

Sandstone rubble in a masonry wall

This red conglomerate stone is a distinctive characteristic of the south-east of the country and was used in the construction of many buildings in, and around, Wexford town

Metamorphic stone

Metamorphic stones were formed from other stones which went through a change brought about by heat and/or pressure. For building purposes, the most common metamorphic stones are marble and slate.

Marble is limestone metamorphosed and it is used for decorative purposes. Marbles do not weather well in northern Europe and so, were mostly used inside buildings. Many stones described as Irish marble are not in fact true marble. Connemara marble is a type of serpentinite and is green in colour while black Kilkenny marble is a limestone that can take a high polish.

While more commonly found as a roofing material, slate was also used as a cladding material for exposed walls, gables and chimney stacks. The slates found in slate-hanging are often rare examples of native Irish slates which, because of their small sizes, were often replaced on roofs by Welsh slates

Slate is mudstone metamorphosed by lateral pressure and heat. It has been used since mediaeval times in Ireland to clad the roofs of prestigious buildings. The best known slate type is probably Welsh slate, which was exported extensively to Ireland from the eighteenth century onwards. Although there were many varieties of Welsh slate used on Irish buildings, the most common variety is Blue Bangor.

Native Irish slates were also used and came from quarries such as those at Portroe, County Tipperary, Ahenny on the Tipperary-Kilkenny border, Valentia, County Kerry and Convoy, County Donegal. Roof slate is no longer quarried at these sites.

BRICK

The use of brick in construction in Ireland dates from the sixteenth century. It was generally made locally where suitable clays were found. Brick was also imported, particularly in the eighteenth and nineteenth centuries. There are few intact early brick buildings remaining, but brick is often found in ruined masonry structures in elements such as chimneys, door and window openings and vaulting. For further information, see *Bricks – a guide to the repair of historic brickwork* in this Advice Series.

Fine early handmade bricks together with limestone dressings were used in the construction of Palace Anne, County Cork built in 1714

MORTAR

Mortar is the substance used to bind stones together in masonry construction. Lime mortar in historic buildings is made from lime putty and sand, sometimes with additives known as pozzolans, to assist the mortar to set. Brick dust was commonly used as a pozzolan. A variety of other additives and inclusions were also used such as crushed stone, shells, charcoal, blood and linseed oil. Animal hair, such as cow and goat hair, was often added to renders to minimise cracking. Lime putty was made by burning limestone in a kiln, and the remains of hundreds of lime kilns throughout the country bear witness to the tradition. Lime mortar was also used as external render and internal plaster.

Clay mortars made from local clays, sometimes with added lime, were also used mainly in vernacular buildings. The clay mortar was also applied to the face of a wall as a daub to bring rough field stone to a level surface.

During the twentieth century, particularly in the second half, the use of lime for building was almost completely abandoned in favour of cement-based materials. However, in recent years, problems caused by the use of cement in repairing traditional buildings began to emerge. Cement is harder and denser than most building stones and when it is used, causes problems, including stone decay and cracking. Cement-based repairs may be encountered by the owners of ruins. A fuller discussion of mortars follows in later sections.

Types of masonry construction

Stone used in historic masonry construction was generally either quarried locally or, in certain parts of the country where available, collected as field stone. Stone suitable for fine carving was often imported or transported to the site by river.

The quality of locally available stone, and its workability, dictated how it was laid, its style and appearance, and whether a building was rendered or not.

Since most historic buildings have been adapted, repaired and altered in their lifetime, ruined structures may often contain a combination of the masonry types described below. Stone, being a valuable material, was often reused from adjacent ruined structures, and it is not uncommon to find carved or worked stones from an earlier structure built into the walls of a ruin or into boundary walls.

A wall showing surviving traces of external lime render. Most rubble-walled buildings were originally rendered to keep rainwater from penetrating into the interior

Even buildings which have stood as ruins for centuries can retain traces of internal plaster. These are most often found in sheltered parts of the building such as here on the intrados of an archway

Earlier carved stone was often reused in the construction of later masonry walls. In this example an opening was blocked up using parts of a disassembled column from the adjacent ruined friary

DRYSTONE WALLING

Drystone walls were built without mortar or any other binding material. Using available stone and reflecting local styles and traditions, these walls are beautiful and distinctive features of the local landscape, particularly in the west of the country. The tradition of drystone walling in Ireland goes back about five thousand years with the earliest known examples to be seen in the collapsed walls at the Céide Fields in County Mayo. Widespread in counties west of the Shannon, some strikingly beautiful drystone work may be found in field boundary walls in the Burren and on the Aran Islands.

Drystone walls show regional variations and styles, dictated by the type and size of field stone locally available (Top and middle images courtesy of Patrick McAfee)

RANDOM RUBBLE

Uncoursed random rubble

Uncoursed or random rubble stone was very widely used in churches, castles, monasteries, town defences and boundary walls. Such walls may incorporate elements of horizontal coursing but not along the full length of the wall. Among the main advantages of uncoursed rubble was that it could accommodate a greater variety of stone sizes. Sometimes through-stones were incorporated to tie a wall together but these were usually randomly located. Uncoursed random rubble walling also has greater inherent strength than coursed rubble which could be more easily overturned by a horizontal force.

Uncoursed random rubble. Note the large quoin stones at the corners which give strength to the structure

Coursed random rubble

This type of construction, where smaller stones are combined with larger to bring the random rubble to a horizontal line or course at regular vertical intervals, was also used; it became more common from the nineteenth century onwards. One of the advantages of this method was that a wall could be tied together using through-stones at regular intervals.

Coursed random rubble

Double skin wall construction

Double skin walls, consisting of an inner and outer stone face, were commonly used in historic structures. The two faces were often tied together with through stones. Mediaeval walls of double skin construction rarely used this detail but it became more common from the eighteenth century onwards. The centre of the wall was filled with a rubble core of smaller stones, lime mortar and sometimes clay.

CYCLOPEAN

Cyclopean masonry was a type of construction originally found in Mycenaean Greece, the last phase of the Bronze Age in ancient Greece. It was built with huge limestone boulders, fitted together with minimal clearance between adjacent stones and no use of mortar. In Ireland, the term is applied to masonry walls where the stones are often edge bedded and not of great thickness. The style is early and can be seen in small pre-Norman churches and also in later churches that retain earlier masonry. A variation of cyclopean masonry called polygonal masonry can be seen in some Gothic Revival buildings of the nineteenth century. Polygonal masonry was formed of close-fitting, many-sided stone blocks which were generally not laid in horizontal courses.

ASHLAR

Ashlar consists of finely dressed stones cut square and having the same height within each course. It was used in the construction of more formal buildings such as civic buildings, churches and country houses. Ashlar joints are often quite fine, which makes repointing difficult.

This image highlights the contrast between the finely worked stonework of the ashlar piers, with narrow joints between the stones, and the coursed rubble walling that was used to block up the former gateway between the piers

Snecked masonry, sometimes called random ashlar, was built in discontinuous courses with small stones or snecks introduced to break courses. The stones required cutting but offered economy in material as smaller stones could be used. This style of construction was first used in the nineteenth century and is distinctly Victorian

Large blocks of cyclopean masonry can be seen to the lower courses of this wall

Features of masonry walls

A number of features may be found in a masonry wall. These include:

BATTER

A batter is an intentional splaying of the base of a wall making the wall wider at its base. It was a construction technique designed to increase the stability of a masonry wall and was used particularly in tall buildings of the mediaeval period such as round towers and tower houses. The inclusion of a base batter may also be seen in some nineteenth-century Gothic Revival buildings which imitated features of mediaeval buildings.

TOOLING

All dressed or worked stone bears the marks of the mason's tools, which give the stone a distinctive finish and sometimes can be used to date the work. There are many types of finishes: rough, smooth and textured and with or without drafted margins - a term for a method of finishing the edges of the individual stones.

There are some distinctive features to watch out for. Diagonal tooling may be seen on earlier mediaeval buildings and brings the face of the stone to a fairly smooth finish with fine comb marks running at an upward angle from the bed. Punched finishes can be a feature of later mediaeval stonework.

MASONS' MARKS

Sometimes the mason cut his own 'mark' on a stone to sign it and also perhaps as a way of keeping a tally of his work. Such marks can be very useful in dating masonry and in identifying the movement of stonemasons between different building projects. Marking on stone may be faint and may only be visible in certain light.

A mason's mark carved into a finished piece of stone. This is a simple mark of a type common in the thirteenth century. Later masons' marks were more elaborate and often in relief. Note the diagonal tooling

In strong sunlight the marks from the original mason's tools are clearly visible on the surface of the stone (Image courtesy of the Office of Public Works)

PINNINGS

Also known as 'gallets' or 'spalls', pinnings are small pieces of stone, brick or tile pressed into the mortar joints of a wall to reduce the amount of mortar required and thus reduce the likelihood of shrinkage and cracking of the mortar. Pinnings can often be used to give a decorative effect to walling and this is particularly the case with buildings from the Gothic Revival period in the nineteenth century.

The small stones used as pinnings can be easily lost as the mortar holding them in place erodes over time. When repairing a wall, it is important to collect any fallen pinnings and reincorporate them into the joints as they are an essential characteristic of the wall

PUTLOG HOLES

These are holes found at intervals in a wall which were used to secure the horizontal timbers of scaffolding during construction. On completion of the wall, the timbers were generally sawn off flush with the wall surface. Over a long period of time, and particularly in ruined structures, the timber pieces left behind rotted away leaving holes in the wall. Putlog holes are most commonly found in masonry walls of the mediaeval period. Other holes in walls were provided to support the ends of beams and joists and can be used to identify original floor levels or locations of roof elements.

Where holes of these types exist they should never be filled in as they are an important archaeological record of the construction of the building and, in some cases, may even contain surviving remnants of the original timbers, providing an opportunity for dating using dendrochronology. Where there is a nuisance from birds, wire can be fitted to prevent birds landing or nesting.

Putlog holes, literally holes in a masonry wall where timbers were put, are a common feature of mediaeval buildings

WICKER CENTRING/PLANK CENTRING

A feature associated with stone vaults, particularly those found in tower houses of the thirteenth century onwards, is a mortar layer to the underside of the vaulting often bearing the imprint of the construction process that was used to build the vault.

Temporary timber supports, or centring, were put in place to support the masonry arch as it was being constructed. Timber planks were placed over the centring to form the shape of the vault. A bed of mortar was then laid over the planking and the stones of the vault set into it. Once the arch was complete and capable of supporting its own weight, the timber centring and planking were removed from below, leaving the imprint of the planking on the underside of the mortar bed, which was now exposed. This was a favoured method of the Anglo-Normans for constructing vaults and arches.

A more common variation of this construction method, particular to Ireland, was the use of mats of woven willow rods, or wickerwork, in place of timber planking. The wickerwork mats were more flexible and easily adjusted and also saved on the amount of worked timber required in the construction of the vault. In some cases, it would appear that the matting remained in place after construction had finished but generally only the imprint of the wickerwork remains in the mortar. However, in some examples, pieces of the wicker itself may survive. It is important to identify marks of centring and to protect them from damage.

In this example of wicker centring on the underside of a vault, not only the imprint of the wickerwork but also parts of the original wicker matting itself have survived embedded in the mortar layer

EVIDENCE OF SURVIVING DECORATION

Masonry walls should always be carefully examined for evidence of surviving finishes such as plaster or render. Because they decay more quickly than stone, especially where a building has become a ruin, the original plaster or render coatings have often largely vanished together with any decorative finishes that were once applied to them. In the past, this led those who studied ancient ruins mistakenly to believe that early stone buildings were not rendered and that they were largely undecorated. However a careful study of the more sheltered parts of a masonry ruin will often reveal valuable evidence of early renders and plasters sometimes with painted decoration.

From the late sixteenth century onwards, the use of renders became more decorative, in some cases using contrasting finishes to imitate cut stone detailing. A contrast could be formed between the coarseness of a roughcast render on wall surfaces and a finer, smooth render, sometimes darkened using additives such as charcoal. In other examples, smooth render was ruled-and-lined to appear as ashlar facing.

Rare surviving painted decoration on mediaeval plasterwork can be seen on this part-concealed pilaster (Image courtesy of the Office of Public Works)

Conservation principles

The primary aim of conservation is to prolong the life of something of value, and to do so in a way that protects what is valuable about it. The built heritage enriches our lives and provides a connection with, and a means of understanding our shared past. Ruined structures, while they may be seen as having little or no economic value in themselves, in fact have incalculable value as tangible historical records of those who have gone before us, of the lives they lived and of their aspirations and achievements. Each structure provides unique and irreplaceable evidence of the past and should be passed on to future generations with that evidence intact.

As each historic structure is unique, each requires an individual assessment of its significance, its condition and a solution to the particular conservation issues that have arisen. The conservation of ruins requires highly specialised skills in all aspects of the works. Expert advice is needed in assessing the extent of works required, designing and specifying those works and overseeing the project on site. Skills are also required of the craftworkers, stonemasons and others who carry out the works to the ruin.

An aim of good conservation is that there should be minimum intervention into the historic fabric of a structure. Conservation works should do as much as necessary, yet as little as possible to the structure to ensure its future. This means that elements should be repaired rather than replaced. Conjectural reconstruction of any part of the structure should be avoided and only undertaken where there is good reason and where the works can be based on reliable documentary or other evidence. Appreciation is needed of all the various phases of construction. Later additions or alterations may be of equal, or in some cases more, interest than the original built fabric.

CARRYING OUT MAINTENANCE OR REPAIR WORKS

> Do use the experts - get independent advice from the right people

> Do repair the parts of the structure that need it - do not replace them unless they can no longer do the job they were designed to do

> Do make sure the right materials and repair techniques are used and that even the smallest changes made to the structure are done well

> Do use techniques that can be easily reversed or undone. This allows for any unforeseen problems to be corrected in future without damage to the special qualities of the structure

> Do establish and understand the reasons for failure before undertaking repairs

> Do record all repair works

> Don't overdo it – only do as much work to the structure as is necessary, and as little as possible

> Don't look at problems in isolation – consider them in the context of the structure as a whole

> Don't use architectural salvage from elsewhere unless certain that the taking of the materials has not caused the destruction of other old buildings or been the result of theft

Legal protection of historic ruins

It is important to establish at an early stage whether a ruin is protected by legislation and what types of notifications, permissions and/or consents it may be necessary to obtain before undertaking any works. This section is intended as guidance only and is not a legal interpretation of the National Monuments Acts 1930-2004, the Planning and Development Acts 2000-2006 nor the Wildlife Acts 1976-2010.

NATIONAL MONUMENTS ACTS 1930-2004

A historic ruin may be protected under the National Monuments Acts in one or more ways as follows:

a) By reason of being a national monument in the ownership or guardianship of the Minister for the Environment, Heritage and Local Government or a local authority or subject to a preservation order;

b) As a monument entered in the Register of Historic Monuments;

c) As a monument entered in the Record of Monuments and Places.

In respect of monuments to which (a) applies, the written consent of the Minister for Environment, Heritage and Local Government is required for any structural interference or ground disturbance. In respect of monuments to which (b) and (c) apply, two months' notice in writing must be given to the Minister of any proposed works at or in relation to the monument. Breach of these requirements is an offence. It will generally be found that at least one of the above applies to any ruined mediaeval structure such as a tower house, castle, town wall or a church in an old graveyard. The Record of Monuments and Places (RMP) is the most widely applying provision of the National Monuments Acts. It comprises a list of recorded monuments and accompanying maps on which such monuments are shown for each county. It can be consulted in county libraries and main local authority offices. The National Monuments Section of the Department of the Environment, Heritage and Local Government will advise on the protection applying to any particular monument under the National Monuments Acts.

PLANNING AND DEVELOPMENT ACTS 2000-2006

Alternatively, or in addition, a historic ruin may be protected under the Planning and Development Acts by being included in the Record of Protected Structures (RPS) of a particular planning authority or by being located within an Architectural Conservation Area (ACA). Where a building is a protected structure (or has been proposed for protection) or is located within an ACA, the usual exemptions from requirements for planning permission may not apply. In the case of a protected structure any works which would materially affect its character will require planning permission. Legal protection also extends to other structures and features within the curtilage of a protected structure such as outbuildings, boundary walls, paving, railings and the like. In an ACA, any works to the exterior of a building which would affect the character of the area also require planning permission. Owners and occupiers of protected structures have a responsibility to maintain their buildings and not to damage them or allow them to fall into decay through neglect.

A notice is sent to every owner and occupier of a protected structure when the building first becomes protected. The Record of Protected Structures can be consulted in the development plan for the area. If a building is a protected structure, or if it is located in an ACA, the planning authority will be able to advise what this means for a particular property.

The owners or occupiers of a protected structure are entitled to ask the planning authority in writing to issue a declaration which will give guidance on identifying works that would, or would not, require planning permission. Maintenance and repair works, if carried out in line with good conservation practice and the guidance contained within this booklet, may not require planning permission. If an owner or occupier is in any doubt about particular proposed works, the architectural conservation officer in the relevant local authority should be consulted. However, if the structure is also included in the RMP, notification under the National Monuments Acts is always required, notwithstanding an exemption from planning permission.

For general advice on planning issues relating to architectural heritage, a publication entitled *Architectural Heritage Protection Guidelines for Planning Authorities* (2004) is available from the Government Publications Sales Office or can be downloaded from www.environ.ie.

WILDLIFE ACTS 1976-2010

Under the Wildlife Acts it is illegal to destroy (whether by cutting, burning, grubbing up or spraying) vegetation on uncultivated land during the bird-nesting season, that is between March 1st and August 31st in any year. While it may not be illegal to cut back ivy or other vegetation growing on a wall or other built structure during this season, best practice should avoid doing so if at all possible. Consultation with the National Parks and Wildlife Service (NPWS) of the Department of the Environment, Heritage and Local Government is recommended and may assist in decision making. While the unchecked growth of ivy can cause serious problems to a masonry ruin, its benefit to wildlife, in particular to bees and birds, is immense and therefore where ivy is not causing problems it should be left alone.

Nesting birds and/or roosting bats are often associated with ruins and in certain circumstances it can be illegal to disturb them. While nests are commonly recognised on the external parts of buildings and bat roosts are often associated with intact roofs, both bats and birds can be found in small cavities in stonework such as exist in ruined masonry structures. If there is any concern that nesting birds or roosting bats may be present where vegetation is about to be cut back or removed; in a structure about to be repaired; or in trees due for felling, advice should be sought from the NPWS.

SAFETY, HEALTH AND WELFARE AT WORK ACT 2005

Ruined structures are, by their nature, potentially dangerous and unstable. When commissioning works to a ruin, the owner or custodian of the structure should be aware of the requirements of the Safety, Health and Welfare at Work Act 2005 and the Safety Health and Welfare at Work (Construction) Regulations 2006. The duties of owners/clients, contractors and relevant professionals are mandatory under this Act and its accompanying Regulations. Helpful guidance is provided on the website of the Health and Safety Authority www.hsa.ie.

2. Approach to Maintenance and Repair

The most important task for an owner or custodian of a ruined masonry structure is to observe it over time, developing a habit of regular inspection, keeping a note of changes and carrying out simple maintenance and preventive tasks. If this is done, the need for repair is seen early. Many structures may be sound, and simple measures will ensure that expensive repairs can be avoided.

Ruins differ from other buildings such as dwellings or workplaces that are used on a daily basis. They are often remotely sited and may not be visited for months on end. This makes them much more vulnerable than other structures. In most cases, structures fall into a ruined state gradually and the rate of deterioration is slow. There are exceptions, and attention may be drawn to a structure by a collapse of masonry, sometimes during a storm. However, collapse generally comes at the end of a slow process of decay. At higher levels, mortar loss, plants and self-seeded trees may not be obvious when viewed from ground level. Vegetation, growth of nettles and rough surfaces underfoot may inhibit casual inspection.

Recording and assessment by the owner or custodian

Before starting, it is important to learn as much as possible about the particular structure. What is its history? How has it changed over time? Later alterations are important and provide evidence that the building has been cared for and adapted over the years with each generation adding its own layer to a unique history. What is its condition now and how can the structure be cared for into the future? What are the responsibilities of the owner or custodian?

The owner or custodian should:

> Be informed about the legal status of the structure and what notifications, permissions or consents are needed for any works
> Be informed about the historical background
> Get to know the building as it stands
> Keep a file on the building
> Get the right advice
> Start a programme of preventive maintenance

LEGAL STATUS

The owner or custodian should first establish the legal status of the structure. The local authority will be able to answer queries regarding a building included on the Record of Protected Structures, such as what are the owner's responsibilities and any permissions which will be required under the Planning and Development Acts.

For structures and sites included in the Record of Monuments and Places under the National Monuments Acts, the National Monuments Service of the Department of the Environment, Heritage and Local Government will advise on the notifications and/or consents required under that legislation.

HISTORICAL BACKGROUND

There are many sources of information available to assist an owner or custodian to discover the history and development of a ruined structure. County libraries are the most immediate source and will be able to point the owner in other directions such as to the Archaeological Survey of Ireland (ASI), the National Inventory of Architectural Heritage (NIAH), Irish Architectural Archive (IAA), National Library collections (including photographic archives, such as the Lawrence Collection), the National Archives and local history societies. The County Library may also have copies of the historic maps prepared by the Ordnance Survey which are useful as an initial dating guide. There are also useful drawing collections, such as those of early antiquarians including Petrie, Grose, Beranger and Du Noyer, in major public institutions; several of these drawing collections have been reprinted in facsimile. Many records are now available online and there are some excellent guidebooks to sources for local history.

Families who have had a building in their ownership for generations will have detailed knowledge and may have documents or photographs of family occasions, with buildings visible in the background, which can provide useful information on the earlier form and condition of the structure. There may be other photographs available locally. Where a building has been altered, an experienced eye can sometimes date the alterations from examination of the types of masonry and techniques used.

Moore Hall, County Mayo. This historic house, constructed by wine merchant George Moore in 1792, is attributed to the architect John Roberts. The house was later the home of the writer George Augustus Moore. It was subject to an arson attack in 1923 and has stood as a ruin ever since. The top photograph shows the house as it stood in the late-nineteenth century when it was photographed for the Lawrence Collection. The middle photograph of the ruin was taken in the 1950s while the bottom photograph shows the ruin in 2006 (Lawrence Collection image courtesy of the National Library of Ireland)

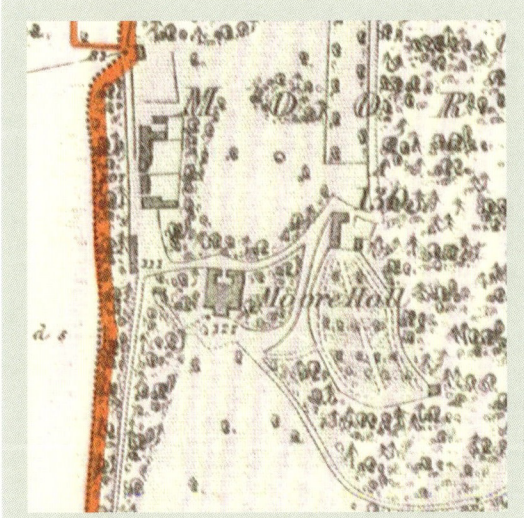

Old maps can be very useful sources of information. This unpublished map of Moore Hall and its demesne dating from the 1860s clearly shows not only the shape of the main house but also the existence and location of outbuildings, a walled garden, paths and planted features of the parkland associated with the house

GETTING TO KNOW THE BUILDING

The owner or custodian should make regular inspections to check the condition of the structure and make a point of doing this after a storm. The extent of vegetation growth in or around the ruin should be regularly assessed and measures taken to ensure it is kept under control.

It is advisable to undertake a detailed inspection annually, taking photographs from specific viewpoints which will be repeated, giving an excellent record of the building and any changes. It would be best if this were done on a specific day each year when conditions might be similar. A private owner could choose a specific date. A local authority might schedule an inspection of all monuments in its care for a specific week. Comparison with the record of the previous visit will quickly show signs of deterioration. Photographs taken after a period of rain can show where water is penetrating the structure. The information collected can prove invaluable in noting changes over time and will be very useful when a professional advisor is engaged for a project.

KEEPING RECORDS FOR THE BUILDING

It is recommended that the owner or custodian open a file on the building and keep a record of all information collected. Copies of historical records, maps, photographs and drawings should be kept along with reports, photographs and notes of inspections by the owner and by the professional advisor. This information should be collected together, for example, in a ring binder. A record should also be kept of any repairs and maintenance carried out, as the owner's involvement in itself becomes part of the history of a ruin.

Checklist for owners or custodians

> Establish the legal status of the structure and the types of consents, permissions or notifications required for any works
> Use established sources to collect as much information as possible about the ruin and its site
> Make a dossier on the ruin which will incorporate all available information
> Establish a pattern of regular observation. Inspect and photograph on a given day each year if possible and record the date
> Recognise when to get advice and what type of advice is required
> Look at the structure with the professional advisor and decide what is appropriate for the particular problem
> Check for evidence of the presence of wildlife particularly nesting birds and roosting bats
> Leave lichens in place
> Try to identify other types of vegetation. There may be interesting historic plants associated with the structure
> Remove seedlings or saplings, but only where this can be easily done without risk of destabilising the masonry
> Leave any loose fragments where they are, recording their location by photograph if necessary
> Pointing of walls should only be undertaken following expert advice: this type of work needs considerable experience and expertise and may require permission and/or consent
> Use experienced and skilled craftspeople
> Record any changes by means of photographs and notes
> Do not take any risks when examining the structure

Getting the right advice

The repair of a masonry ruined structure is highly specialised work and the best and most effective time to get advice is from the outset. Good advice is never wasted and may save money in the longer term. Professional advice is best, as architects, engineers and archaeologists must keep abreast of current legislation and practice and are trained to take a broader view. Look for a building professional, an architect or engineer with the relevant conservation skills and experience. The Royal Institute of the Architects of Ireland (RIAI) has an accreditation system for architects trained in building conservation and can provide a list of those architects that have the necessary skills and training for your project. The National Monuments Service of the Department of the Environment, Heritage and Local Government has a list of licensed archaeologists. The architectural conservation officer in the local authority can provide general advice and may have information on suitable professionals, craft workers and suppliers in your area.

Consideration should be given to commissioning a preliminary report early on which will point the way forward. It will prove useful in seeking grant assistance and avoiding costly mistakes, so leaving scarce resources to be spent on the building itself. In most cases, a short initial report outlining the significance of the building, giving an appraisal of its condition and making repair recommendations should be sufficient. If the building is protected under the National Monuments Acts or is particularly complex, it is often more efficient for a building professional to work in a small team, seeking appropriate input where necessary from specialist colleagues such as archaeologists and architectural historians. In some cases, a multi-disciplinary team may be a requirement of a consent or permission. Each member can contribute by working to their strengths in their special area of interest. The owner should get a concise, competent and wide-ranging report which does not end up costing the entire repair budget.

Remember that it makes sense to engage someone from the outset who will not simply write a report, but is qualified to see a project through to the end and to advise on technical issues as well as on choosing contractors, seeking tenders and looking after work on site where necessary. In many cases, the advisor may be needed for an initial consultation only but may be called upon at a later stage if the need arises.

What to expect from a professional advisor

The following is a general list of the services which can be provided by a qualified, experienced conservation advisor. Taking into account the size, importance and condition of the structure, and the requirements of statutory authorities, the owner and advisor can agree on which of the services are necessary and agree a fee for the scope of services to be provided.

> Desk study of documentary evidence concerning the history of the site
> Site visit for preliminary examination (to verify the historic account, make an assessment of overall condition and identify issues relating to public safety and the safety of those who may have to work on the ruin)
> Advice and recommendations on the level of recording necessary
> Licensed archaeological investigative or test excavations
> Liaising with the appropriate statutory authorities (for example the planning authority and/or the Department of the Environment, Heritage and Local Government) to determine their requirements for the conservation of the site
> Photographic record of the site as found
> Detailed survey of the exposed ruin
> Identification of issues relating to trees, shrubs, ivy and general vegetation
> Wildlife habitats, consultation with wildlife specialists and the National Parks and Wildlife Service
> Identification of species of plant for historic context
> Recommendations for initial control of overgrowth to facilitate a detailed examination of the ruin including issues of timing of work to avoid disturbance of wildlife, especially protected species
> Analysis of historic mortars
> Report incorporating the results of recording and evaluation; recommendations for temporary works, safety barriers, scaffolding and access; evaluation of the significance of the ruin and its setting in the landscape or townscape
> Report and method statement for the conservation of the ruin, its setting and the natural environment of the site based on the principles set out earlier
> Preparation of a grant application for the works
> Advice on seeking tenders and preparation of tender documentation
> Administration of a contract for works

Level of recording

The proper recording of a structure and site, prior to any works, is a fundamental principle of good conservation and its importance should not be underestimated. Depending on the significance and complexity of the structure, the existence of any surviving timber, decoration, painting or plaster, and the level of repair and/or intervention required, a decision will be required on what type of survey and recording is the most appropriate. Professional advice will be needed in this regard. Recording and survey may include any of the following or a combination of methods listed below from the simplest to the most sophisticated.

> Study of relevant original drawings, historic images and papers
> Rough sketches with annotation supplemented with photographs
> Preliminary survey and freehand drawings, to scale, which can be supplemented with photographs
> Measured survey and detailed drawings of plans, sections and elevations
> 3D laser survey and rectified photography or other form of digital recording

While reports will vary in type and content depending on the size, complexity and significance of the ruin, most reports will normally consist of:

> A written account with a short historical account, an assessment of the significance, a condition assessment and recommendations
> Drawings or sketch layouts
> Copies of historic maps
> Photographs

Preventive maintenance

Regular maintenance and correct repair practices will extend the life of any structure, keep it from falling further into decay and can be considered part of the day-to-day responsibility of all owners. It is less damaging and less expensive to carry out regular works of small-scale maintenance than to postpone any action until a major intervention is the only way of securing the structure's conservation. The following points should be borne in mind:

> Uncontrolled vegetation growth is probably the single greatest cause of damage to a masonry ruin
> Plants can seed themselves in voids in masonry and on open wall tops. Routine observation and removal of seedlings and saplings can prevent the necessity for extensive repairs at a later stage

An open wall-top will provide a home for wind-borne and bird-borne seeds which thrive in their new home

Ivy will displace stones as its stems thicken and can became part of the structure, holding up the stones it has dislodged

> If the ivy growth is light, keep it under control by cutting through the stems. New growth will almost certainly come through with small shoots whose berries can provide food and shelter for small birds
> Lichens are mostly welcome on a wall and should not be simply cleaned off. Some lichen types can cause damage to some stones and, in special cases, should be carefully removed following expert advice

Lichens will form on most stones and can indicate the presence of clean air

> Check trees nearby for overhanging branches or contact in windy conditions
> Note the extent of vegetation. The removal of vegetation without proper consideration can increase the risk of masonry collapse. Heavy ivy cover or other vegetation may have deep roots, including aerial roots, that is, root systems that have penetrated the wall at all levels. Killing off such vegetation can make the ruin unstable. Removing vegetation such as ivy may require notification if the structure is a recorded monument or consent if a national monument
> Avoid cutting back or spraying vegetation between March 1st and August 31st of any year if there is a likelihood of the presence of nesting birds. Bear in mind that disturbing wildlife such as nesting birds or roosting bats may be illegal and, if in doubt about how to proceed, seek expert advice

Graves close to a ruined church wall can damage the masonry foundation itself and undermine the earth supporting the foundation

> Note changes in ground conditions and look for possible reasons. Watch out for drainage works on adjoining land which may alter existing ground conditions

> Look for signs of cracking, subsidence or bulging walls

> When examining a ruin, look out for any loose masonry fragments. Record their location but do not remove them without good reason and only following expert advice

> Ruined structures can present considerable safety issues. If there is loose masonry, ladders should not be used and observation should be carried out at a safe distance. A pair of binoculars is very useful in inspecting a structure safely, particularly the higher and more inaccessible parts. It is possible to have a photographic survey carried out from a remotely controlled camera on a high telescopic boom mounted on a rough-terrain vehicle. This may be an economic alternative to an access hoist if there is concern regarding the condition of the structure at high level

> End the practice of burials close to walls. In ruined church sites with graveyards there is a serious risk of destabilising a ruin by digging graves too close to the walls. Earlier grave digging may have removed most of the foundation support for the wall and one new excavation may be sufficient to cause the wall to collapse without warning

Grant aid

Conservation grants are available for the conservation and repair of protected structures and are administered by the local authorities. You should contact the relevant one for guidance on whether the works you are planning are eligible for a grant and, if so, how to apply. These grants are not available for routine maintenance works, alterations, or improvements nor can grants be given for works already carried out. The type of works must fit within the schedule of priorities set out by the local authority. In order for works to qualify for these grants, they must be carried out in line with good conservation practice. Repair work following the guidance set out in this booklet should be considered as satisfying this requirement.

The Civic Structures Conservation Grant Scheme, administered by the Department of the Environment, Heritage and Local Government, provides grants for the restoration and conservation of buildings in civic ownership or occupation and generally open to the public and which are deemed to be of considerable architectural merit. Applications are accepted from local authorities, civic trusts and other 'not-for-profit' organisations. Further information and application forms are available to download from www.environ.ie.

The Heritage Council provides grants for conservation planning and works. Further information on current schemes and application processes is available from the Council's website: www.heritagecouncil.ie.

Tax incentives are available under Section 482 of the Taxes Consolidation Act 1997 for expenditure incurred on the repair, maintenance, or restoration of certain buildings of significant scientific, historical, architectural or aesthetic interest or gardens of significant horticultural, scientific, historical, architectural, or aesthetic interest. The building or garden must receive a determination from the Revenue Commissioners who must be satisfied that there is reasonable public access to the property. Application forms can be obtained from the Heritage Policy Unit, Department of the Environment, Heritage, and Local Government.

3. Common Defects Found in Ruined Structures

In a ruined structure, masonry behaves differently from that in a roofed, intact building. A ruin will probably have been originally protected by a roof, its greatest defence against the elements. Exposed masonry allows water to penetrate and this leads to a loss of mortar in the joints between the stones and in the wall core. Most of us will assume that a ruin has been there for centuries and will survive in the same state for many centuries more. Nothing could be further from the truth. The most robust structures, open to the weather, will deteriorate progressively and eventually collapse if not maintained.

It is important to carry out periodic maintenance, to monitor the behaviour of the structure, to note any changes and defects, and to recognise when the masonry has reached a stage when it needs to be repaired. To do so is part of an owner's responsibility to keep a protected structure from endangerment whether from neglect or damage. However, regardless of the legal status of the structure, it is good practice to carry out regular maintenance checks of a masonry ruin of any type and to keep unwanted vegetation under control to prevent a potentially dangerous situation emerging wherein the structure may become destabilised.

Damage from vegetation

Uncontrolled vegetation is a major factor in the deterioration, decay and collapse of masonry ruins. Nature requires no assistance in taking over sites of masonry ruins where there is no maintenance regime in place. However, there may be restrictions on the timing and method of works to control or remove vegetation if there is a likelihood of disturbing wildlife such as nesting birds and roosting bats.

Many ancient sites were planted with trees in the nineteenth century to make them more picturesque. The trees are now fully mature and entering their decay phase, so that even large trees fairly remote from the ruin can present a risk of substantial damage should the tree or branches fall.

Mature trees adjacent to a ruin present a challenge. While they may add greatly to the beauty, character and natural heritage of a site, they can cause damage by undermining the foundations with their roots, or by falling branches

It is important to get expert advice when assessing the impact of trees on a ruined structure. Not all species of tree have root systems of a type that damage foundations. Where substantial trees surround a monument, they should be included in an overall appraisal of the site, with the assessment of the trees being carried out by a suitably qualified tree specialist. If the tree in question is known to be used by nesting birds or roosting bats, the advice of a competent ecologist should be sought before undertaking any works.

THE IMPACT OF VEGETATION ON MASONRY

There are three main issues relating to vegetation attaching to, or in close proximity to, masonry ruins:

a) Direct damage done by ivy and/or saplings growing on the structure. The roots of the vegetation grow into the joints in the masonry and, as they grow over time, lift and open the joints so that, ultimately, the masonry depends on the root system to hold it together. Roots can also cause mortar to disintegrate and lose its strength and it becomes liable to being washed out of the joints.

Ivy can become established within a masonry wall, growing unchecked through the joints and re-emerging in other parts of the structure

b) The weight of heavy vegetation on a masonry structure can cause damage and make the structure vulnerable to wind damage. The effect increases greatly in wet and stormy conditions.

c) Vegetation affects soil moisture, particularly in heavy clay soils, by extracting moisture from the soil. Dense ground cover improves water run-off, which also has an effect on soil moisture. The destruction of vegetation by spraying ground cover with herbicide and killing off ivy or small trees close to the structure can result in a rapid rise in soil moisture. Significant changes in soil moisture can destabilise the structure so the removal of vegetation should proceed in a very controlled manner following careful assessment of all the factors and risks involved.

Vegetation which is allowed to grow unchecked can rapidly overwhelm a structure and accelerate its decay

Foundation failures

Foundation failures in ruins are rare where the ground around the ruin has been left undisturbed. This is so because poor foundations generally fail early in the life of a building or, if not then, soon after its abandonment. There are, however, examples where there have been a number of interventions to correct foundation failures while the building was in use. These may have been reasonably effective at the time but slow deterioration has continued after the site is abandoned. Fortified structures such as tower houses and their enclosing walls may have been deliberately undermined at some stage in an attempt to destroy them. Where the attempt was not successful, the later backfilling of the excavation may not have provided adequate support leading to continuous slow failure. In the case of ruined churches, the most common cause of foundation failure is burials close to, beside, or even under the walls of the church.

The continuous weakening of the ground near a ruin may lead to a very substantial collapse often blamed on a storm rather than on the weakening of the foundations. Walls on an east/west axis, which are founded on clay soils, can be affected by differential movement caused by the slow and continuous rotation of the foundation towards the south. This is caused by the differences in soil moisture on either side of the wall. On the northern side, where there is almost continuous shade, soil moisture varies very little, whereas on the southern side the soil moisture fluctuates due to heating and cooling and also due to evaporation on the south face of the wall, leading to shrinkage in the clay beneath.

Defects in walls

WATER PENETRATION OF WALL TOPS

Water penetration of wall tops is a common and significant cause of deterioration in masonry structures. Wall tops may originally have been covered either by a roof or, where exposed, capped with stonework, both of which provided weather protection. In the case of ruins, wall tops may be broken and uneven, creating traps for rainwater which seeps down through the core of the wall so that the wall not only gets wet on both faces but also internally. A careful evaluation of the wall top is needed before deciding on remedial action. In some instances, a grassed cap has developed on the wall, which has been colonised by short, drought-resistant, grass varieties which, in the absence of other invasive growth, may best be left undisturbed. In other cases, where there is no protection or where the growth is invasive, a very careful programme for sealing the wall top and encouraging rainwater to drain away may be necessary.

Wall tops are particularly vulnerable to being colonised by grass and seedlings. While some of these plants may be relatively harmless, others such as ivy and saplings should not be allowed to take root

Tapering cracks can result from subsidence. As the crack widens the smaller stones fall out or drop inside the wall creating a wedge effect and increasing the damage

CRACKING

Large tapering cracks, which are wider at the top than at ground level, are usually due to foundation failure, either in the cracked section of wall itself or in an attached wall. Cracks with little or no variation in width from top to bottom but where the wall face on one side of the crack has moved relative to the other can be due to rotational foundation failure. Cracks at wall openings usually indicate a problem with a lintel or an arch. Projecting masonry on one side of a crack relative to the other may be the remains of the springing of an arch or a corbelled gutter or wall walk.

Vertical cracks can be more common in walls where few long facing stones were used. Longer stones create a stronger bond than smaller stones

It is important to establish a programme of measuring and monitoring the rate (if any) of cracking in a wall. Some cracks may have been there for a very long time and show no signs of further movement; others will register a progressive acceleration in movement. Monitoring may be done in a number of ways, the simplest being a mortar dab across a small crack.

PARTIAL COLLAPSE

High, freestanding walls with little or no lateral support can be blown over in high winds. This is particularly the case with exposed gables and chimneystacks. The thickness of random masonry walls generally provides sufficient resistance to being overturned or blown over. High gables can be vulnerable, as they slowly deteriorate with a loss of mortar in a number of joints, developing a rocking motion in high winds which, if not arrested, will lead to failure. More commonly, partial collapse is the result of excessive growth of vegetation which increases the impact of wind on the wall. Walls with a south-facing aspect can develop a slowly increasing lean to the south which, ultimately, leads to collapse of the weaker portions of the wall.

Partial collapse of this church tower occurred suddenly, taking down a portion of the nave wall

LOSS OF WALL CORE

The most common method of rubble masonry construction was to build, in stages, an outside and inside face with substantial stones mortared in place, and progressively to fill the gap between them with smaller stones and general stone and mortar waste from the construction process. This filling between the two masonry faces is called the wall core. In some structures, the wall core has plenty of mortar, and in others, little, but quite frequently there is no mortar at all. This may be because the amount of mortar was inadequate from the time of construction or that water lodging within the wall has softened the mortar allowing it to be washed out. So when one face of the wall is lost, the wall core often falls out, leaving a substantial area of wall in an unstable condition. Loss of wall core can also occur where the thickness of the wall has not allowed the mortar to set properly.

Wall with its outer face lost exposing the construction of the wall core

LEANING WALLS

Leaning walls or parts of walls are common in ruined structures, and it can often appear that the wall is about to collapse. Where a wall or part of a wall is leaning, it is essential as a first step to monitor it to establish whether the movement occurred many years ago and the wall has settled, or if the leaning is progressive, possibly due to changed circumstances, and the wall is likely to reach a point when it will topple.

A leaning structure which has stood for many years in this position

BULGING AREAS OF WALL

Bulges in a wall face may arise from a variety of causes. The movement or rotation of foundations is a common cause of wall bulging. So also is the growth of ivy or tree roots within the core of the wall, which may push the wall face outwards and the bulging remains even after the tree or ivy has been removed. The bulge may also be the result of defective building in the first instance, where an adequate number of through stones was not used to create a stable wall face. In clay-bonded walls, the gradual loss of strength of the clay bedding and subsequent washout results in a re-arrangement of the masonry of the wall face which can result in bulging. This is less usual in walls built using only lime mortar.

FAILING ARCHES

Arches in masonry ruins fall into four general categories:

a) Arches spanning substantial openings in walls or across large spaces, such as chancel arches and crossing towers or an arcade between the nave and side aisles in church buildings. There may also be substantial arched vaults in tower houses, supporting towers and turrets and in mansions. This category of arches is very vulnerable in masonry ruins. These arches are dependent for their survival on the presence of other masonry to resist the horizontal thrust from the springing of an individual arch or of the last arch in an arcade. They have much less lateral stability than a wall and they are very vulnerable to the erosion of mortar between the voussoirs (tapered stones of the arch). In such cases, the most obvious sign of deterioration is the spreading of the arch, i.e. its flattening, at the springing points, or the dropping or displacement of voussoirs as the mortar in the joints erodes. The stabilising of substantial arches is a difficult undertaking which will require specialist professional input as well as the services of a very experienced contractor.

One arch on this double bell-cote has already failed. The other is on the point of collapse due to loss of mortar through erosion and exposure

b) Arches to carry the walls over window openings and door openings. These arches are generally smaller in span and may have secondary masonry such as window tracery within the spanned opening. These arches can suffer from all the defects listed for major arches above. Their shorter spans make them appear to be much more stable. However, if a series of closely-spaced lancet lights (as in a mediaeval church chancel) loses lateral support to the springing of the final arch, then a progressive failure can occur through the entire set of windows. Smaller openings tend not to have shaped voussoirs and are therefore much more heavily dependent on the mortar between the stones to keep the shape of the arch. Loss of mortar is quite often the cause of failure in these relatively narrow arches.

Arch over a window. Note the damage done by the improper removal of ivy which pulled out the smaller packing stones and has left the arch vulnerable

Arch over a door opening. This arch is also vulnerable because the wall over the arch has been lost

Relieving arch over a flat door lintel

d) Flat arches made of brick were common in the eighteenth and nineteenth centuries for window and door openings. Flat arches made of stone, with secret joggles connecting the stonework, were also in use from an earlier period. Flat arches are subject to the same causes of deterioration as described for other arches.

A flat arch constructed of brick with a limestone keystone

c) Relieving arches are also used over doors and windows, but used where the top of the opening is squared up with the flat lintel below the arch. They are also found at the base of walls in areas of poor ground conditions. Relieving arches appear to be more stable because they are embedded in the wall face. They are only protected, however, from the failures mentioned above by virtue of the flat lintel which squares the opening below and supports the infill masonry between the lintel and the relieving arch. The loss of the lintel, commonly of timber, on the inner face of the wall can leave the arch more vulnerable, since relieving arches are frequently not built to the same exacting standard as arches that are required to stand on their own.

FAILING LINTELS

Lintels are flat, horizontal structural supports for the masonry above an opening. Lintels come in a variety of materials and forms depending on the age of the building. They may be made of stone spanning over small openings such as windows and doors, or more substantial cut and dressed stones spanning between columns, for example in a classical colonnade. Single-piece stone lintels have very little resistance to bending and, for the most part, depend on an arching effect in well-mortared masonry above the lintel to reduce the load on the lintel itself. Stone lintels have a limited ability to cope with structural movement which redistributes the load from the masonry overhead onto the lintel. The most common form of failure in stone lintels is straight-through cracking. They are often found in this condition with the masonry overhead surviving because a new natural arch develops within the masonry. There is, however, very little to stop such lintels from falling out. Timber was widely used in lintels right up to the twentieth century and iron and steel came into more general use in the nineteenth century. Clearly, timber in masonry ruins is subject to rot, and exposed steel and ironwork to rust.

The condition of the surrounding masonry is important in the survival of timber lintels. If the wall top or pointing is poor, water will lodge on top of the timber lintel and cause decay

Lintel failure in this case was probably the result of the loss of mortar and stone from the wall above the lintel which damaged the natural self-arching tendency of the masonry placing excessive loading onto the lintel

Stone lintel over a narrow opening. The thinness of the lintel, particularly at the right-hand end, makes it vulnerable to cracking

DECAYED AND WEATHERED MORTAR

Ruined masonry structures may have been built with lime/sand mortar, with clay mortar or with a combination of the two. The natural progression with a pure lime mortar is that it becomes harder with time as it gradually absorbs carbon dioxide from the air and slowly returns to its original status – limestone. This cycle can be disrupted by plant growth which creates acidic conditions resulting in the decay of the mortar. Decayed plant material accumulating on top of the wall also creates acidic conditions and that material can be slowly washed down into the core of the wall and, in time, destroy the mortar.

Decayed/weathered lime mortar

Delaminating stone, bedded vertically

Clay-mortared walls are entirely dependent on dry conditions to retain their strength. The use of clay mortar without the supplementary use of lime mortar pointing, and possibly also a lime render, would be most unusual in any substantial building and, therefore, the loss of pointing mortar, render or weather capping of the wall top inevitably leads to the collapse of these walls.

Mortar and renders on ruined structures inevitably are worn away by action of the weather. The walls are exposed on two sides; therefore, they become saturated in wet weather and, with a change to a heavy frost, the formation of ice within the joints results in the erosion of the mortar.

DECAYED STONE OR BRICK

Sandstones, oolitic limestones, brick and terracotta are all very vulnerable to weather damage, by either frost action or concentrated rain action. Rubble masonry was rarely intended to be exposed and when the building was in use would usually have been protected by lime render.

When the render is lost, soft stones not intended for exposure become vulnerable to weathering, particularly stones which have been incorrectly laid. Squared sandstone rubble was frequently sawn at right angles to the bedding plane of the stone; thus, when placed in the building the bedding plane is in a vertical position. The same is true of calp limestone found in buildings in the Dublin area. Substantial erosion of stonework can take place by frost action, which pushes off large flakes of stone, since stone is weakest along its bedding plane.

Oolitic limestones were widely used for decorative work because of their ease of carving. They are, however, very porous and weather poorly if badly detailed or if the weathering detail has failed to protect the stone, as in the case of many ruins. Oolitic limestones were mainly imported from England, but also sometimes from Northern France and sometimes designated by their origins. Dundry, Bath and Portland stone are examples.

Carved sandstone, showing weathering. Sandstone was widely used for carving because it was relatively soft and easily worked. Some high crosses and Romanesque detailing were carved in sandstone

Brick and terracotta also have high porosity and are frequently only weather-resistant on one face. Many bricks were poorly fired, especially those made before the widespread use of commercial temperature-controlled kilns in the nineteenth century. These bricks can soften due to continuous moisture retention and erode back from the exposed face due to frost action.

'ROBBED OUT' MASONRY

Stone was always expensive to quarry and it was much simpler and cheaper to use stone from a masonry ruin than to quarry new stone. Masonry ruins have always been subject to the plundering of their stonework. Cut stonework, either in squared rubble or ashlar, together with moulded or sculpted masonry, was highly prized and frequently robbed from masonry ruins. Most commonly, the moulded stonework around windows and doors and the squared and dressed stonework of quoins were robbed out, creating substantial areas of weakness in the remaining masonry.

It must be remembered that the unauthorised removal of stone from any recorded monument or protected structure is illegal and subject to prosecution and penalties.

Stone has been robbed from this section of a town wall and probably used to build other masonry structures nearby

Decisions on whether and how to replace robbed-out masonry need to be carefully considered. In conservation terms, masonry should not be replaced unless it is structurally necessary. Masonry ruins can often be more structurally stable than they appear.

Replacement should only be considered where it has been established that the ruin is structurally unsafe or unstable or as part of a well-considered conservation project. It is advisable to have the structure assessed by a competent building professional with experience in the repair of historic masonry.

EMBEDDED IRON CRAMPS AND ARMATURES

Iron cramps were frequently used in construction, from the eighteenth century onwards, to connect masonry pieces together within the thickness of the wall and sometimes also to tie heavy moulded masonry back to the main structural stonework of the wall. This is particularly the case with deep projecting cornices. Armatures, or iron frameworks, were also used in the construction of slender columns or for supporting bosses and tracery. Both forms of ironwork were intended to be concealed in dry locations and were often seated or encased in lead. However, in the case of ruins, the ironwork inevitably got wet or damp resulting in rusting and expansion and thus damaging the stonework which it was intended to support. Where the damage is destabilising the masonry, it may be necessary to dismantle the masonry to remove cramps and replace them with non-ferrous material or with stainless steel, while at the same time repairing damaged stonework. Armatures were generally used in delicate moulded masonry and rusting and expansion can result in very significant damage. It can be difficult to remove armatures while avoiding damage to the masonry in the process.

DEFECTS CAUSED BY EARLIER INAPPROPRIATE REPAIRS

Masonry walls which were re-pointed or repaired using cement mortar may, over time, give rise to problems. Dense cement pointing, which is non-porous, traps moisture within the wall. In some instances, the hard pointing mortar causes the softer stone to crack. Sometimes, the cement pointing may fall out, showing the original lime mortar behind. While it is generally advisable to remove cement pointing where it is causing damage, the removal works may cause further damage to the stone and a decision may be made after careful assessment that it is less damaging to leave it. Some earlier repairs or repointing may have used a lime mortar gauged with cement which, while containing cement, may not be as damaging as fully cement mortars. Such gauged mortars can be left in place unless there is evidence that they are causing damage. In any case, power tools should never be used to remove cement pointing, unless in the hands of specialist conservators, because of the potential to damage the face of the stonework. This also applies to cement-based wall-capping which is much more likely to crack and allow water penetration in concentrated areas.

Many historic masonry structures were originally plastered, using a lime plaster. Lime plaster weathers and wears away with time, and the walls may have been re-plastered at some stage using cement render. As with cement pointing, cement plaster is hard and dense and, with constant freezing and thawing, will crack allowing water into the wall. This water becomes trapped within the walls and causes damage to the masonry.

Structural insertions such as concrete ring beams or cement-rich grout introduce areas of unacceptable inflexibility into structures that are used to accommodating minor movement, leading to further stress and potentially serious damage.

Strap pointing is a style of pointing which must be avoided on historic masonry structures. Unfortunately, it is quite a common intervention and was often used where the original joints were very fine. Unsightly and damaging to the stone, the pointing is proud of the stone surface and traps water on the ledges formed. It also partially adheres to the face of the stone and is almost always cement-based

4. Planning for Repair

The primary objective in repairing a ruin is to stabilise the structure and prevent further deterioration. Many ruins will present a range of different problems as a result of the loss of a roof such as vegetation growth, vandalism and exposure to the weather. Hence, any one structure may require a variety of repair techniques.

In approaching the repair of ruins, there are a few important general principles to be followed:

> Make sure the reasons for failure are fully understood before undertaking repairs
> Only repair the parts of the structure that can no longer do the job they were designed to do
> Only do as much work to the structure as is necessary, and as little as possible
> Consider problems in the context of the structure as a whole
> Even minor repairs may require archaeological input where the structure is a recorded monument
> Many types of repair will require temporary works such as propping or scaffolding
> Make sure the right materials and repair techniques are used, and that even the smallest changes made to the structure are done well
> If possible, use techniques that can be easily reversed or undone. This allows unforeseen problems to be corrected in future without damage to the special qualities of the structure
> Record the state of the structure before and after all repair works
> Avoid the use of salvaged material from elsewhere unless it is clear that the taking of the materials has not caused the destruction of other old buildings or been the result of theft
> Avoid using coatings and chemical treatments on masonry and/or plaster, except limewash where appropriate

Principal materials used in repair

STONE

In many cases, where stone for repair is required, it may be available on site. If a simple collapse is recent, the original stone may be lying on the ground exposed. If the collapse is historic or if stone is buried or covered in undergrowth, recovery may be more complex. Searching for and extracting the stone will require licensed archaeological excavation in some cases. Where stone has been robbed, and has left a structural weakness, replacement stone should be of the same geological type and of similar appearance - from the same quarry, if possible. Dressed stone however, may not have been sourced locally. The true colour of the in-situ stone may only be judged from a non-weathered surface. In some cases, it may be necessary to carefully knock a small chip off a concealed face of the existing stone to form an accurate judgment of the original colour. The replacement stone may look quite bright initially but will weather in time.

BRICK

Depending on the date of construction, brickwork to be repaired will consist either of earlier, handmade brick of irregular size and shape, or more recent mass-produced brick. In either case it may be difficult to source a modern replacement brick for repair. The lost brick may not be available on site as it may have failed through cracking or disintegration. The use of salvaged material may be the only option. For further information, see *Bricks – a guide to the repair of historic brickwork* in this Advice Series.

LIME MORTAR

The use of lime mortars in the conservation of historic ruins is important for three basic reasons:

> To replicate as closely as possible the original materials used in the construction

> To create a mortar that is less hard than the stone in the construction, so that expansion of the mortar, due to frost or other action, will result in failure of the mortar rather than damage to the stone

> To provide 'breathability' in the wall. Lime mortar is the most vapour-porous of all mortars. It allows moisture in the structure to evaporate through the mortar joints, so avoiding damage to the stone by salt build-up or frost damage if this becomes the main route of evaporation.

Ordinary Portland Cement cannot meet the above three criteria, and therefore should not be used in the conservation of historic buildings and structures.

An analysis of the existing mortars may be necessary to guide the choice of aggregate for the repair mortar and also as a record of the historic mortar present in the structure prior to works commencing. Professional, scientific analysis of historic mortars can be undertaken by a number of commercial companies in Ireland.

SAND

Sand for lime mortar should be clean, sharp sand well-graded from coarse to fine, usually 2mm or 3mm to 150 microns, with the highest proportion of particles occurring at the mid-range. Many existing mortars will have some larger particles of 5mm or more. The coarser particles help to dull the brightness of the fresh mortar as they are exposed when the mortar is tamped back. The colour of the finest or smallest particles can also affect the colour of the mortar. The quality of the sand is important and the proportion of lime to sand is a function of the sand grading.

LIME

The basic forms of lime in use today are little changed from those known throughout the period from which most of our ruined structures survive. Although some of our limestones could theoretically produce lime with feebly hydraulic properties, Irish lime nowadays is made from very pure limestone and is typically non-hydraulic.

Lime is produced by burning limestone, which releases carbon dioxide and leaves calcium oxide, known also as quicklime, behind. The slaking of the quicklime produces calcium hydroxide, which is available as an ingredient of mortar, either in the form of lime putty or dry hydrate powder, known as 'bag lime.' This is the form of lime produced in Ireland from our own limestone. It is pure calcium hydroxide and when used in mortar, hardens - initially by drying and over a longer period by carbonation (the absorption of carbon dioxide from the air to form calcite). Pozzolans have from earliest times been added to mortars made with putty limes to accelerate the set, and by tradition were finely ground pumice, or finely ground brick.

The manufacture of building limes is now more closely controlled than it was in the past, and a Europe-wide standard gives a uniform product. Almost all of the high calcium, that is non-hydraulic, limes from Ireland and the UK are produced for industrial and agricultural use, and only a tiny part is used as putty lime or hydrate powder in the building industry.

Naturally hydraulic lime

Some non-Irish limestones naturally contain silicates which, when burned produce what is known as hydraulic lime. The material is available as a dry hydrate powder consisting of calcium hydroxide and reactive silicates. Hydraulic limes harden, initially by chemical reaction between the ingredients in the presence of water, and in the longer term through the absorption of carbon dioxide. Hydraulic limes produce an early set by chemical reaction and, once set, are not affected by additional water such as might arise by wetting of the structure by rain. The resultant hardened hydraulic lime mortar consists of calcite, calcium silicate hydrates and sand. While some hydraulic limes are produced in the UK, most are imported from continental Europe, mainly from France. Naturally hydraulic limes are available in four grades or strengths which are categorised according to the strength they achieve in tests after 28 days: NHL 1, 2, 3.5 and 5 N/mm². NHL 5 is the strongest and mostly too strong for the repair of historic structures.

The choice of lime

Historically, lime used in Ireland was non-hydraulic and used with or without pozzolans. The choice of the type of lime to be used today, whether non-hydraulic or with varying degrees of hydraulicity, should be based on the analysis of the historic mortar which it replaces, as well as the performance requirements for the mortar, together with the environmental conditions and exposure level of the site. The best advice is to use a mortar compatible with that analysed. A word of caution - analysis may show a hydraulic set, generally the product of the inclusion of impurities in the slaking process, and this is not to be taken necessarily as an indication of the presence of hydraulic lime. The analysis may also be based on samples of inconsistent or degraded mortar.

Lime putty, also called 'fat lime', should be used where possible. It is a softer material producing a weaker, more flexible mortar. Because of the type of setting process, it requires greater protection for a longer period while it carbonates, so work must be well planned. Winter weather conditions are not favourable to external use.

Soft or porous sandstone masonry or clamp-fired brick require weak mortars. In less exposed locations, good quality mortars made with putty lime, if well made and appropriately used, may be adequate but will require weather protection until a hard skin has developed. The work will also require protection from wetting from above and through the core, since the long term stability of lime putty mortars depends on the mortar taking up carbon dioxide from the air to induce hardness in the mortar. Alternative options in such situations can involve the use of a carefully selected pozzolan to accelerate the set of the putty lime mortar or the careful gauging with a weak hydraulic lime mortar. Where the stones are sound and durable and in exposed conditions, the use of a weak to moderately hydraulic lime mortar may be necessary.

Mix proportions for the various types of lime mortars and the grading of the sand and coarse material to be incorporated will depend on the analysis of existing mortar samples and a professional evaluation of the other issues set out above.

Working with lime

Lime should generally be purchased in slaked form as the process of slaking is highly dangerous. Quicklime, delivered for slaking, is a highly hydroscopic material and hence when combined with water produces a very powerful reaction. When not in use, quicklime should be stored in sealed containers and in a shed within an area deliberately set aside for such use. Only craft workers skilled and experienced in such work should be permitted to handle quicklime and to undertake on-site slaking.

Slaked lime, as calcium hydroxide, delivered as either putty or dry-hydrate, is highly caustic and poses a potential health risk. Working without gloves, leading to its direct contact with the skin, can cause the skin to dehydrate and even burn. Lime should be used with great care in accordance with all current safety regulations. Gloves and barrier creams are essential skin protection, as are goggles for the eyes. Like cement, dry-hydrated non-hydraulic or hydraulic limes may be inhaled during mortar preparation, so this should always be undertaken outside and face masks worn; the same precautions should also be followed when using powdered pigments. In addition, once dry-hydrated limes are mixed with water, as with lime putty, it is critically important to avoid direct contact with the skin.

Lime mixes set more slowly than cement, and slow setting allows the material to achieve maximum strength. Lime is more susceptible to weather conditions, and sets more slowly in winter. It is also vulnerable to damage by frost. Work should generally be avoided between late autumn and early spring. Sun and wind in turn may over-accelerate the rate of setting. If repair work is necessary in unsuitable weather conditions, protection should be provided by draping hessian over the masonry. In dry sunny weather, the hessian may need to be sprayed with water to keep it damp.

Protective hessian on masonry work is far more effective than plastic sheeting and can be reused many times

CLAY MORTARS

The term 'clay' in this instance is used as a descriptive, rather than a soil classification, term. A typical clay mortar will consist of fine sand, silt and clay (given here in descending order of particle size), and may also sometimes contain fibre in the form of straw and dung. Lime is sometimes an ingredient. The proportions of the constituents in the mortar may vary widely from location to location with sand making up between 30% and 50% of the mortar. Clay mortar should be used to match existing material, so the mix will depend on analysis of the original mortar. The proportions of the constituents of the historic mortar can be readily checked by mixing a sample in a jar of water until it has fully broken down. The mixture is then thoroughly shaken and allowed to settle. Different sized particles of sand, silt and clay settle at different rates and, when final settlement has taken place, the ratio of the constituents can be measured on the side of the jar.

The preparation of clay for use in mortar is made, ideally, by exposure over a period of a year so that the action of frost and rain breaks down the clay and silt particles. The quantity of prepared clay to be mixed with the sand can then be kept to the minimum needed to coat the sand particles and bind them together. Clay shrinks very substantially on drying, and therefore trial mixes should be made to achieve the optimum result.

Clay mortar is traditionally mixed by hand, but there is no reason why it cannot be mixed in a mechanical mixer provided that the clay element is well and truly broken down and no clumps are introduced into the mixer. Water must be added very sparingly and only the bare minimum to produce workability being added, otherwise the clay and silt particles will expand and then shrink substantially on drying.

Clay mortared walls in ruins will always require special protection to shed water from the top of the wall to prevent softening of the mortar from inside. The wall-top protection will have to be done with lime mortar. Clay mortared walls were traditionally either pointed with lime mortar and/or rendered, or in lesser quality buildings given several coats of limewash.

Temporary works

PROPPING OR TEMPORARY SUPPORTS

If a potential collapse has been identified by, for example, a slippage of masonry or cracking, temporary propping may be necessary to secure part of the structure from collapse. It may also be necessary to provide temporary supports to allow safe access for close examination, recording and assessment of the structure. The form of temporary prop needs to be assessed, having regard to the length of time the prop is likely to be in place, the extent of public access and the history of vandalism on the site. A prop installed on a temporary basis must be adequate to arrest the deterioration in the structure, and therefore, the removal of a prop as the result of vandalism or other inappropriate behaviour may result in a serious fall of masonry. Props should only have a restraining effect and not create any upward pressure.

Timber props with folding wedges are suitable for short-term use. The timber, however, will shrink and it may be necessary to tighten up the wedges on a routine basis.

Timber bracing in the vertical plane to secure a window opening where a collapse outwards was imminent. The temporary work consisted of planks bolted through the opening

Props which may need to be in place for a number of years while a programme of conservation work is carried out are best made of modern masonry such as concrete blocks or bricks set in weak lime mortar, with a separating membrane used to facilitate later removal. The placing of temporary props needs to be carefully considered so that they do not impede the conservation work, thus requiring premature removal.

When propping is being considered, the ground conditions need to be assessed. There may be sub-surface archaeology which will be affected, or the ground may be uneven, with mounds and cavities concealing fallen masonry or old burials. Heavy machinery should not be used in the vicinity of the ruin because of the damage it may cause to the ground underneath.

Providing access

Scaffolding is frequently needed to facilitate a full assessment of the ruin, and it is essential that the weight of the scaffolding is well distributed, particularly in the case of graveyards. Scaffolding should not be attached to the structure, instead raking supports may be needed to keep the scaffolding stable. The design of scaffolding is governed by Codes of Practice for wind-loading, vertical loading and transfer loads from structures and must be done by a suitably qualified person. Hoists or 'cherry-pickers' can be a useful and economic way to provide access for carrying out an assessment of a structure. However, the impact of such machinery on the surrounding ground and any sub-surface archaeology should be carefully considered, and it may not be appropriate in certain conditions, such as in graveyards.

FENCING AND HOARDING

Temporary fencing will be required during conservation work, as is normal on work sites. The risk of damage to artefacts near the ruined structure is increased while work is in progress; this is particularly the case in relation to graveyards. The location of temporary facilities needs to be carefully considered, such as containers for materials storage and welfare facilities, including toilet facilities for those working on the project. The access route to the ruin needs to be carefully chosen and defined to avoid accidental damage, particularly to headstones and other monuments and sub-surface archaeology. Headstones and other architectural decorative masonry in the immediate work area will need to be protected by substantial plywood boxing against the risk of falling masonry or workers' tools and mortar splashes.

A clear sign warns of conservation works in progress. The temporary works to the wall in the background include scaffolding poles to provide restraint. The concrete blockwork on the right hand side provides temporary propping pending reinstatement of the wall core

Graveyard with headstones protected by plywood casings in advance of conservation works to the adjacent structure

Loose carved stone and other artefacts

Loose stones from the structure or its fittings or fixtures may also be found on the site. They may include carved or dressed stones, effigies, crosses, or cross-inscribed slabs. If these are likely to be affected by the repair works, they should be photographed and their location carefully recorded. If the site is included in the Record of Monuments and Places, stones should not be moved without the prior approval of the National Monuments Service. With approval they may be stored in a safe place while consideration is given as to their treatment when the work is completed. It is preferable to retain such masonry in or near the location where it was found. There are a number of options as to how loose stones should be fixed to prevent theft and vandalism, and protected to prevent further decay. Sometimes, when the stone is of particular quality, it may be necessary to consider providing some form of shelter on site. However, this should be carefully considered as there is a danger of creating a micro-climate within the shelter, leading to further deterioration of the stone. The advice of a stone conservator should be sought in such a case.

Carved stones lying in vegetation

Dealing with vegetation

Trees and plants around ruins can be aesthetically pleasing and are also immensely beneficial to wildlife, particularly to bees and birds. However, uncontrolled vegetation can cause serious structural damage leading to collapse. The first rule is to keep it under control. The second rule is when the vegetation is out of control and well established in or close to the walls, it is usually preferable that nothing should be done until the situation is assessed and resources are available. The intervention should be well-planned. The assessment should include the cover or canopy of vegetation and the wind-loading caused, the stems and root systems, access and other safety issues, and the timing of the interventions.

IVY REMOVAL

Ivy removal, if not correctly carried out, can cause a significant risk to a ruined structure and may have an adverse effect on wildlife, including protected species. Under the Wildlife Acts, it is illegal to cut, burn or spray scrub during the bird nesting season and this applies to any vegetation growing around or within a ruin. While it may not technically be illegal to cut back ivy or other vegetation growing on a wall or other structure during this season, best practice would avoid doing so if at all possible. Consultation with the National Parks and Wildlife Service is recommended on how best to proceed.

As a first step, the ivy canopy should be reduced without interfering with the root system. This will allow for a closer inspection of the masonry and reduce the risk of wind damage to the structure. The use of power tools should be avoided as they may cause damage to the surface of stonework. However, unless a comprehensive treatment programme is being put in place, severe cutting back to allow a close inspection should be avoided as this can encourage a vigorous re-growth, leaving the plant stronger than ever.

The top layer of vegetation has been sprayed and killed off pending its removal. The treatment allows for assessment and recording of the masonry

Treatment of ivy to kill the growth should be carried out some time before the works start. The stems should be cut through above ground level by hand saw and a section removed, if the stems are accessible. It may be appropriate to spray the foliage with an approved systemic herbicide to kill an extensive root system, some of which may be living in or feeding from mortar in the walls.

Ivy removal should only proceed when a programme of conservation work to stabilise the structure in conjunction with the removal has been approved and is ready to start. Ivy should never be pulled off while alive – this can cause structural instability or collapse of masonry. It should always be allowed to die back before careful removal by hand - never by rope. The ivy can be cut back but removal of the ivy roots is best left until the masons undertaking the conservation works are on site. The structure may then be secured in sequence with the roots removal.

Removal of an ivy canopy may reveal deeply embedded roots and stems which have penetrated the masonry to such an extent that it has become unstable. If a wall has become so destabilised that the mortar is no longer binding the original stones together, dismantling the masonry may have to be considered.

TREE REMOVAL

Trees around a ruin, if they have been growing for many years, may be an intrinsic part of the ruin in its setting. They may also provide a habitat for wildlife. However, tree roots and overhanging branches in close proximity to the structure may pose a threat to the stability of the walls and foundations.

The roots of trees that have been allowed to grow up too close to a wall can threaten the stability of a structure

Before any decision is made to remove a mature tree, it is important to get appropriate expert advice. A tree felling licence from the Department of Agriculture, Fisheries and Food may be required. Advice should also be sought on any planning, archaeological and other issues in relation to the removal of trees on the sites of protected structures or of recorded monuments. If the tree is known to be used by nesting birds or roosting bats, the advice of a competent ecologist and the National Parks and Wildlife Service should be sought. The removal of healthy trees may need to be undertaken over a number of seasons depending on the nature of the subsoil. In heavy clay soils, it is best to proceed by reducing the tree canopy slowly over a number of seasons to allow the soil moisture to stabilise before the tree is finally removed and its roots killed.

The least damaging way of doing this is usually to cut branches back above the base; if they are of substantial size, they should be cut down in sections that can be manually carried off the site. The use of chain saws in the pruning and removal of trees close to a building or wall - even by experienced personnel - should be avoided as they can cause damage to the masonry.

Removal of roots needs archaeological supervision as the roots are likely to have penetrated into archaeological material. In some cases, it may not be necessary to remove the stump. Further growth should be prevented by the most appropriate means; it may be acceptable to poison the root system through the truncated base with an ecologically acceptable herbicide, but expert advice should always be sought on this. As an alternative to the use of poison, closely set copper nails hammered into the truncated base is thought to be an effective way of preventing further growth.

The use of biocides should generally be avoided and only used where absolutely necessary by competent persons.

TIMING OF TREATMENTS

The timing of treatments should take account of the bird-nesting season (from March 1st to August 31st) during which it is generally illegal to cut, burn or spray vegetation on uncultivated land. Work on trees is best carried out between October and February, when there is minimal foliage. Old trees should be regularly monitored and if necessary lopped or removed before damage is caused by wind-throw.

The removal of substantial vegetation needs to be timed to periods when soil moisture is naturally stable. The situation will vary according to the species of vegetation involved.

The timing of the removal and treatment of vegetation is also related to building conservation work which, for the most part, is best carried out during the period April to October, unless the access scaffolding is fully enclosed and winterised.

5. Repair Techniques

The defects listed in Chapter 3 may arise from a number of causes, which need expert diagnosis. Many of the problems encountered in a ruined structure will be complex and may require the application of a combination of the techniques listed below.

Underpinning foundations

Most of the foundations beneath historic ruined structures would be considered inadequate in modern engineering terms. Yet although they appear inadequate, the structure has remained standing, often for several hundred years. This highlights the importance of engaging an engineer with experience of historic structures particularly traditionally-constructed masonry.

Underpinning is the method of restoring solidity to the foundations of walls. It can involve the extension of the foundation down to firmer bearing strata, widening the foundation to reduce the bearing pressure, or the installation of piles which extend down to good bearing strata and are then connected to the existing wall foundation.

The extension of the foundation down to better bearing strata is sometimes referred to as 'conventional underpinning.' In this method, short excavations are made under the wall down to the new bearing strata where a new foundation of concrete is cast. The wall is then re-supported from that new foundation with either a concrete or masonry wall to connect with the original foundation.

The excavations take place in a carefully predetermined sequence so as not to destabilise the wall. The second method is to create a larger foundation by casting a much wider new concrete foundation just beneath the existing foundation. This has the advantage of requiring less excavation but has the disadvantage in archaeological or cultural terms of creating a greater area of disturbance in the vicinity of the wall. As with the conventional method, the sequence of sections must be carefully planned so as not to destabilise the wall.

Piling techniques may be applicable in some instances. Other methods involve the injection of grout or resin into the soil beneath the wall to improve the bearing capacity of the soil. These techniques are highly specialised.

Wall cappings

Where good quality coping stones exist and have been dislodged it may be necessary only to clean the wall tops, clear them of vegetation and re-set the coping stones.

Where most of the coping stones survive and just a few are missing, there is a good case to be made for introducing new material to replicate the originals, since the evidence of the original design survives on site. Where wall tops are broken and uneven, the temptation to level them off and provide a cap of modern materials should be resisted. It is generally not necessary and a hard, clean line on the skyline of a ruined building may detract from its appearance. In addition, the loss of historic fabric in levelling a wall top will almost always be considered unacceptable.

Careful cleaning of the wall top to eliminate vegetation and the removal of accumulated humus is essential in the first instance. It may be necessary to re-set stonework which has been dislodged by vegetation. The wall top should then be mortared with a stronger, hydraulic lime mortar to shed water from the surface so that it does not penetrate into the wall. This flaunching is laid to a fall but this should not be obvious when viewed from ground level. The surface of the wall should be finished so that water is shed easily over the edge while not, at the same time, creating a full mortar cap on the wall. Where the wall top is visible and easily accessible, a 'soft-capping' technique may be used; here, grassed sods are used on wall tops to absorb rainfall and protect the wall surface. While soft capping has ecological benefits, it is advisable to maintain it under observation and not encourage future problems of uncontrolled vegetation growth, in particular, woody plants.

Flaunching, a sloping area of mortar, has been applied to the raking surface and also to the upper surface of this wall on which a mat of vegetation is beginning to grow

Stitching

Stitching refers to the tying together of cracked or damaged areas of masonry using modern materials. One of the most common applications of stitching is to re-tie a wall on either side of a crack using steel ties laid into joints at intervals. The crack can then be pointed up with mortar. Failed stone lintels can be repaired in place using stainless steel rods drilled in unobtrusively and set in epoxy mortar, with the drill hole being concealed by a pellet of the original stone.

Failed lintel with stainless steel rods used to carry out a stitch repair

This form of repair can be used also to strengthen arches in conjunction with wall core grouting, and to tie the outer and inner faces of the walls together to prevent them from separating.

Stitching can also be used to carry out in-situ repairs on fragile masonry elements such as window tracery or detaching moulded stonework.

Rebuilding collapsed masonry

If there has been a recent collapse of part of a plain rubble wall caused by weather or vegetation, and the rubble is lying on the ground, there is no reason not to repair the collapse. The faces of the stones may be identified as they will have no mortar adhering to them. The masonry should be rebuilt in character with the surviving wall in similar style and workmanship, taking account of the style and shape of any pinnings, coursing patterns, stone shapes and the like. If the owner has been keeping a record as discussed earlier, there will be a photographic record that can be used to inform the work. The rebuilding of a historic collapse is more complex. All the material may not be still available and rebuilding may not be structurally necessary. The advice of an experienced conservation professional will be required.

Partial dismantling and rebuilding

Repair of protected structures or monuments should be carried out on the basis of minimum intervention. There will however be occasions when it is not possible to secure an area of masonry without dismantling some stonework. The decision to dismantle and rebuild a wall, part of a wall or other section of masonry should only be made when all other options have been explored, and it is considered that this is the only way to ensure structural stability. Such a decision should only be taken following expert professional advice, and in consultation with the Department of the Environment, Heritage and Local Government in the case of a recorded monument or the planning authority in the case of a protected structure.

Before dismantling, all the stones should be numbered and the section of wall recorded photographically. The numbering of stones should be carried out using a water-soluble paint that can be washed off later without damage to the surface of the stone.

It is never possible to exactly replicate the work and the rhythm of masonry created by a long-dead mason, whose tradition and training would have been quite different to that of the present day. Nonetheless, it is essential not to create disharmony in the rebuilding.

Masonry wall with stones numbered to allow for accurate rebuilding. The strings are stretched across the length of the wall to mark the wall at vertical intervals to ensure joints are kept at original thickness when rebuilding

A repair to a town wall using new stone where the wall had been damaged or robbed. In this case, the new stone was sourced from a quarry 8 km away and, in the course of time, will weather in

The most reliable method of recording for rebuilding is the full-scale drawing of the stonework in place onto a transparent plastic sheet prior to dismantlement. The plastic sheet is mounted in a timber frame, or as many timber frames and plastic sheets as are needed to cover the section of wall, and with a fixed reference system and support system for the frames which will survive the dismantling of the wall.

This system allows for the as-near-as-possible replication of the original masonry. Dismantling and rebuilding alone may not be sufficient to secure the section of wall, and it may be necessary to incorporate modern materials such as stainless steel or bronze to tie sections of the wall together and so avoid a replication of the original cause of deterioration. The rebuilt section of wall is classified as new work and should be recorded as such. There is never a case for introducing architectural or carved details for which there is no evidence.

Repairs to leaning walls

Leaning walls may present a risk of collapse. However, not all leaning walls are in a progressive state of failure, and some may be quite stable. The first step should be to determine the extent to which it is out of vertical alignment along its length and whether it is still moving. The measuring points should be permanently marked and the measurements recorded. The survey procedure should then be repeated at regular intervals. As a general rule, if the centre of gravity of the wall remains within the middle third of the thickness of the wall and if continuous movement has not been detected, it can be assumed to be stable. The middle third rule is only a rough guide, however, and a final decision on stability will require evaluation by a professional structural engineer, experienced in work on historic ruins. The evaluation will require a detailed measurement of the shape of the distorted wall, together with accurate measurements of thickness, an evaluation of the quality of the soils at foundation level and the orientation of the wall in relation to its exposure to wind. There are many methods which can be employed to stabilise such a wall, but the solution should be specific to the particular problem and designed by a suitably experienced engineer.

Repairs to failing arches

Arches may fail through lack of lateral support, through loss or decay of mortar, as a result of stone decay in the voussoirs from frost action, or from a combination of all three. The failing arch needs to be supported in the first instance. If the shape of the arch is not too much distorted, it may be possible to bend a strip of plywood to the shape of the arch, installing timber studwork underneath to support it with folding wedges inserted into any gaps between the formwork and the soffit of the arch.

Where there is evidence of spreading of the arch and obvious movement at its supports, lack of lateral support is the primary cause of failure. The remedial work to restore lateral support should be carefully designed and will be specific to the particular site.

In rubble stone arches, where the failure is due solely to loss of mortar and/or stone decay, and where the stones of the arch may have slipped but are still in place, it is possible when supported to rake out the mortar and re-point the stones, making provision to grout the core of the arch by the insertion of grout tubes. It may be possible to re-set some of the slipped stones without disturbing the adjoining stones, but it is quite acceptable to consolidate the stones in the position in which they are found if they cannot easily be re-set. Once re-pointed, the core of the arch may have to be grouted, as described elsewhere.

Arches made with voussoirs may also deteriorate through loss of mortar. The procedure to be followed is basically the same as described above for rubble masonry arches. However, it is frequently the case that it is not possible to re-set the voussoirs in their original position and the tightness of the joints may make pointing and/or grouting quite difficult. In such cases, it may be necessary to provide extra support in the form of stitching with stainless steel rods. This, however, is a specialist technique and should only be carried out under professional supervision.

Repairs to failing lintels

Stone lintels over narrow openings which have cracked may be repaired by first propping the arch, then using a specialist technique of dowelling with a resin stone adhesive across the crack. Timber lintels, where decayed, may be replaced in an appropriate timber capable of withstanding exposed conditions but with a protective layer of felt bonded to the new timber over and at both ends.

Replacement stones

It is generally not appropriate to introduce new material into a structure. Exceptions to this rule occur where it is necessary to ensure the stability of the ruin, to support important carved or moulded fragments, or to assist in the general interpretation and appreciation of the structure. There is never a case for introducing new material on a conjectural basis. Any new material, particularly moulded or carved stone, should relate directly to existing material of good provenance on the site. Rubble stone buildings, which have had elements of dressed stone robbed out, such as window and door surrounds or quoins, are best retained in that condition by careful stabilisation of the surviving masonry around the openings. Missing facing stone, which needs to be replaced for the stability of the wall, may well be found within the site of the monument, and a good search under archaeological supervision is always worth while before introducing material from outside the site. The local quarry from which the stone was originally sourced will most likely no longer exist, but suitable local stone may become available arising from renovations or demolitions of existing buildings in the locality. Cut and dressed stonework presents a particular problem since the original stone may not have been sourced locally or even within the country. It may, therefore, be neither practical nor possible to source matching stone. Where small quantities of replacement stone are required, such as on verge copings or parapet copings, then an exact colour match may not be necessary as the replacement stones will not be overly intrusive. If the new masonry is required for structural reasons, and there is no clear evidence of the original detail, new stone of similar type and colour may be used, to the same profile, but without any detail.

Where window tracery is missing, and there are structural problems as a result, the introduction of modern materials such as steel fixings set into the mortar joints may be considered appropriate.

Repointing

The first step in repointing is to select and rake out all defective pointing. There is no need to repoint the whole wall if some pointing is sound. The raked-out material should be collected and retained. The raked-out material from rubble work will also contain pinnings or small stones which were set into the face of the mortar in the wider parts of the joints. It is important that these stones be retained and reused, as they give the rubble masonry much of its character and are also necessary for technical reasons as they reduce mortar shrinkage.

A mediaeval wall repointed with pinnings used in the larger joints

Repointing should be preceded by careful cleaning of the joints. All vegetable matter and decayed mortar should be removed and the joint cleaned and dampened before repointing. New mortar is pressed into the joint to a depth of at least one and a half times the height of the masonry joint to ensure the stability of the new pointing mortar.

Samples of the raked-out mortar should be saved for analysis. It may be considered desirable to have a chemical analysis of the lime from the original mortar to determine its properties and whether pozzolan had been added to the mortar to induce a hydraulic set. Hydraulic set means that the lime, water and sand mix hardens by chemical reaction. A sieve analysis of a substantial sample of the mortar is necessary so that the texture of the replacement mortar will match the original as closely as possible. An analysis of the sand and grit may also help to identify the source which, in any event, will always have been relatively local. The lime for the mortar mix will depend on the analysis and on the circumstances. It will be either a lime putty or, in demanding conditions, a naturally hydraulic lime, most usually grade NHL 2 or NHL 3.5, in combination with sand and grit. The sand should be washed, sharp, and graded as close as is practicable to the sand in the original mortar. Grit should be added in the same proportion in which it occurred in the original mortar. The exact ratio of lime to sand and grit will be determined by the grading of the sand being used and the degree of exposure of the wall. The mortar, properly mixed, should be quite 'fat' in texture, but with a well controlled water content so that the mortar when placed remains firm. The mortar needs to be tightly packed into the joint to ensure that it engages with the masonry, and struck off level with the face of the stone. As the mortar is setting, it needs to be tamped or hammered on the face with a stiff bristle brush to finally tighten up the joint and to expose the coarse aggregate in the mortar. Ruins may have broken wall ends which should be finished rough-racked, that is with the exposed ends of the stones left showing in a natural, jagged line with the mortar joints between them roughly pointed.

Grouting rubble masonry

Grouting is a technique for solidifying the core of a wall where the original mortar has decayed or perhaps where the core originally contained insufficient mortar, or none at all. As a result the inside and outside leaves of the wall can take on an independent existence. Specialist firms usually use a cementitious mix and this should be avoided as it may have unforeseen consequences and, if problems arise, the process cannot be reversed. Grouting is a difficult procedure which must be preceded by the consolidation of the exposed masonry, the removal of vegetation, the raking out and cleaning of joints and the repointing of joints. The repointing is most important as the grout has to be contained within the wall. In the raking-out process, some joints are raked right through to the core of the wall in a regular pattern and a plastic grout tube is fitted and held in place by the re-pointing mortar. On completion of the repointing, a lime grout is inserted under pressure of gravity or using a hand-operated grout pump. Grouting continues until the grout appears at the next tube vertically. Both tubes are then temporarily plugged and the sequence continues along the wall and then to the next level, and so on. As the grout hardens the grout tubes can be withdrawn and the joints repointed. Grout, by virtue of its high water content, is not suitable for filling very large voids. These may have to be opened up to allow sufficient access for conventional filling with mortar and small stones. The specific mix and methodology is a matter for specialist advice.

Protection of carved stone and other decorative features

Some ruins may have a carved or worked element such as an inscription or a piece of sculpture set into a wall. It may have become vulnerable to decay and weathering due to the loss of the roof, a string course or a hood moulding. It may be appropriate to introduce a lead strip or slate into the joint above the carved stone to provide alternative protection.

Larger pieces of carved or sculptural work may require more extensive protection to maintain them in situ, and some form of shelter may be considered. However, if the shelter is enclosed, there is a danger that a micro-climate may be created which may further damage the stone. It may also affect the appearance of the monument. The choice to provide this form of protection requires expert conservation advice.

If there is a very significant element of carved stone built into a masonry ruin, and it is vulnerable to destruction by the elements or by vandalism, consideration may be given to removing the carved element to a more secure and stable environment. If this is being considered, the element should preferably be re-located on the same site, near its original location. In some cases, it may be appropriate to incorporate a replica of the original carved element at the location. Again, expert advice will be necessary as well as statutory consents/permissions if the structure has legal protection.

Inscriptions can be subject to weathering and the text eventually lost. It is important to record the inscription and lodge the record in an appropriate location, such as a publicly-accessible archive or a local studies journal.

A ruin may contain fine carved features such as this tomb set into the wall. The condition of the carving should be monitored to ensure that no excessive weathering is taking place

Shelter coats

The application of a number of shelter coats of limewash may offer some protection and a consolidating effect in some cases. Where existing hood mouldings are formed in plaster or Roman cement, a lime-based slurry on the upper surface will protect against weather and aid water run-off. Shelter coats may not always be visually acceptable because they change the colour of the surface to which they are applied. If stone is carved, a shelter coat may obscure fine detail.

Surviving plaster, render or slate-hanging

It is generally not appropriate to remove or replace lost external render or internal plaster on a ruin. What is important, however, is to protect the surviving remains. A fillet of a weak and compatible mortar should be applied to the exposed upper and side broken edges in advance of works to protect the plaster and to prevent water penetration behind it. If the plaster is moving away from the wall, it may be possible to inject a weak lime grout to fill the void. This can be a risky technique that can result in the loss of plaster if incorrectly executed and should only be carried out by an experienced craftsperson. If the plaster contains decorative detail or a painted finish, it should be recorded by drawing and photography prior to any work commencing. If it is sound, a conservation plasterer can take a 'squeeze' or mould from it. As with carved stone, surviving exposed historic plaster should be protected; this can be achieved by lead or slate set into the wall above to protect it from weather.

For surviving patches of slate-hanging, the fillet of mortar can be applied to the upper and side edges, as for plaster, and the exposed top slate pinned back to the wall with a non-ferrous fixing.

Surviving timber or evidence of timber

Occasionally, surviving scraps of timber are to be found in a ruined structure, and where they exist, they are generally of interest and should preferably be left in place undisturbed. It is important that details are recorded. Dating samples of oak timber by dendrochronology is possible if the significance of the building warrants it. This will require the involvement of specialist advice as the extraction of the required sample for dating purposes is prohibited under the National Monuments Acts unless a licence is in place to alter it. The export of a sample abroad (including to Northern Ireland) for dating purposes requires an export licence. Lintel sockets, fixing grounds, joist sockets and putlog or scaffolding holes are the most likely spots to find surviving timber fragments. Wallplate ends or sockets may have survived in gables or the imprint of the wall plate may be visible in a mortar bed at eaves level.

Surviving floors or paving

Historic ground surfaces may have survived in part and should be recorded, consolidated and protected from further damage. Temporary consolidation may be achieved by infilling gaps with a weak mortar and covering during the works. Long-term consolidation and presentation is a matter for expert advice.

Surviving granite paving to an external yard in a ruined mining village

Dealing with emergencies such as collapse

Occasionally a structure will suffer an unexpected collapse or partial collapse. It may be the result of gradual decay which has gone unnoticed or unattended, or may be caused by high winds, a storm or lightning strike. The priority is to ensure the safety of persons, including the general public, by fencing off the area. Some work may be required to make the collapsed structure safe to allow access to assess the damage and the potential for further collapse. It is strongly recommended that the advice of a structural engineer with expertise and experience in dealing with traditional masonry structures is sought.

A decision will have to be taken whether to rebuild the structure or to stabilise the collapsed section. This will depend on the extent of the collapse and the availability of accurate records of the structure before collapse. If the structure is a recorded monument, the National Monuments Service should be consulted or, if a protected structure, the planning authority.

Collapsed structure

Distinguishing between old and new in repair work

While it is an aim of good conservation work to make all repairs visually unobtrusive, it should be possible for the interested observer to distinguish repairs from the original work. When a masonry repair is first completed, the mortar is fresh, the stone may be cleaner and the new work can be disconcertingly bright. However, if the correct materials have been used, in a relatively short time, weather and nature will combine to tone down the new work, and soon a good repair may be indistinguishable to all but the expert eye. As the years go by, the ruin may become more 'illegible' and the original fabric no longer identifiable from the repair work. There is a range of techniques which may be employed on the repair itself in addition to marking the repair on a drawing. Some light punching by the mason of the stone edges at the perimeter of the repair is effective. Another method involves the incorporation of long strips of lead, well embedded in the joint and projecting beyond the face of the wall during the work but bent back flush into the joint at the time of the final pointing. They may be placed at regularly spaced intervals, almost certainly not seen by the general public, but apparent to those with a serious interest in studying the ruin. Where dressed replacement stone is introduced, the year of the repair in small numerals may be cut into the corner of the new stone as a record.

In the past, other techniques of distinguishing between old and new work were used which are no longer considered appropriate. These included separating the new work from the old using strips of building felt. Another method was to recess new stonework from the face of the original work.

A date has been etched into the stone to distinguish the new stone from the original

A method of distinguishing between old and new masonry in repair work was used by the Office of Public Works in the mid-twentieth century and can still be seen on many national monuments today. The technique involved separating the new from the original by the insertion of a strip of felt. This technique is no longer used today

The use of consolidants

The use of chemical consolidants in conservation work is the subject of much debate and the jury is still out on the issue. Their long-term effect in Irish conditions is not known and they do not suit certain stone types. If a structure contains dressed stone in very poor condition, consolidation should be considered only as a last resort and treated with great caution. In considering whether to use consolidants, expert advice should be sought and ideally the issue should be considered by an expert multi-disciplinary team including the relevant statutory authorities.

Stone cleaning

The cleaning of stone for aesthetic reasons alone is not appropriate in a ruined structure. Many cleaning techniques can cause damage to the surface of the stone. However, there may be instances where there are damaging deposits on carved stone or where unsightly graffiti needs to be removed. If this is the case, cleaning should only be carried out by a professional conservator, preceded by trial samples to establish the preferred method. Wire brushes must never be used on historic stonework under any circumstances. Grit-blasting should also be avoided although in some exceptional circumstances the use of small air abrasive tools and fine abrasives may be considered appropriate if used by highly skilled and careful operatives.

Lichen growth on stone

Lichens should only be removed in extreme cases, since their removal will damage the surface of the masonry, particularly in the case of lettered, worked or sculpted stone. Re-colonisation will result in a deeper level of damage. However, some types of lichens are damaging to stonework and expert advice should be sought on their removal. Lichens can be used by experts to help date stonework and, therefore, the need to clean and remove lichens should be very carefully assessed for each individual piece, and other non-invasive techniques for reading and recording the stonework should be tried before embarking on lichen removal.

6. Archaeology

Generally, most repairs should not require any interference with sub-surface archaeological material. However, if a recorded monument requires works such as underpinning, this will raise archaeological issues. The owner or custodian of a ruined structure should be mindful of their obligations in relation to archaeology under National Monuments legislation for all ruins included in the Record of Monuments and Places (RMP).

Archaeological testing and monitoring requirements

The local authority or the National Monuments Service of the Department of the Environment, Heritage and Local Government may request the involvement of an archaeologist during works at the ruin where it is included in the RMP and where there will be archaeological impacts. Archaeological testing may be requested ahead of proposed works on a ruin, especially where rubble clearance, foundation work or examination of original floor surfaces are envisaged, to assess the likely damage (if any) of the proposed works. This includes enabling works such as shoring or scaffolding.

The archaeologist may be asked to monitor the work, to record any archaeological material disturbed and, by assisting in the formulation of the methodology of works, to prevent unnecessary disturbance. When works include clearing stone or soil from the site, the archaeologist will usually be required on site during the entire clearance period. When works involve little or no disturbance of the ground a series of regular visits may be considered appropriate.

Sub-surface archaeology

All material below present ground level at the site of a masonry ruin is of archaeological interest. Even the topsoil may contain important artefacts such as pottery which can assist in tracing the history of the ruin. A licensed archaeologist should be engaged if there is to be any disturbance below present ground level.

Buried building remains include the in-situ fabric of masonry walls, impressions of former timberwork, flooring, construction debris, holes and trenches which formerly supported upright timbers. Other deposits may be equally important. These include domestic debris and silts accumulating in and around a building, the fills of pits and ditches cutting through or cut by the building, and demolition material. All are vital to understanding the date and development of the site.

Disturbed archaeological material will be found where the site has been damaged. This material will add to the understanding of the original site and throw light on the events that damaged the site.

Archaeological investigation revealed the existence of a stone-covered subsurface drain running beneath this mediaeval priory complex (Image courtesy of the Office of Public Works)

Site works

Care should be taken not to disturb archaeological material during site works. This applies to any site clearance, including the removal of topsoil. Where the site is of archaeological importance, work such as widening site entrances, drainage works, erection of hoarding and levelling for temporary site accommodation should only be carried out with the prior agreement of the National Monuments Service and according to their requirements.

Sites such as early mediaeval church sites may extend far beyond the current confines of the church and graveyard. Care should be taken not to damage upstanding earthworks or sub-surface archaeology by allowing the site to be traversed by machinery or other vehicles.

There may be outworks associated with a ruin which should be identified and protected during the course of any conservation works

Removal of tree roots

Even small trees and shrubs can be firmly bound by their roots to material of archaeological interest. Grubbing out roots can cause serious damage, and should only be considered in special circumstances and with specialist advice.

A close examination of the soil caught in the roots of this fallen tree reveals the ribcage of a human skeleton (circled in lower image)

Collapsed masonry

RUBBLE ON SITE

The arrangement and volume of rubble around ruins is significant. In many cases, the best stone will be found to have been 'robbed-out' and removed for use elsewhere, and much of the rubble at the foot of the walls may be clearance from surrounding fields. In other cases, rubble may result from a small partial collapse and be required for repair. Particularly when a building has collapsed suddenly, large fragments of walls, with windows and doors, may lie flat on the ground. Valuable information about the former standing building and the catastrophe that overtook it can be gathered from the archaeological investigation of the rubble. Rubble should not be cleared from ruins except under the direction of an archaeologist.

A partial collapse of the town wall of Youghal, County Cork. Recently-collapsed structures should not be approached until after a structural engineer has deemed it safe to do so

IDENTIFYING ORIGINAL LOCATION OF FALLEN STONES

This requires the combined work of the archaeologist and conservation specialist. Fragments of stonework may be found throughout the site of the masonry ruin by careful ground examination. Covered fragments should be identified with a unique marker and their location carefully mapped. Collapsed material likewise needs to be carefully sorted under archaeological supervision and examined for any signs which may help to identify their original location. The context of the fallen masonry is important. If sufficient masonry survives in situ to indicate the shape of the section before collapse, it may be possible to reuse the fallen stones to carry out a repair at that location. In very rare cases, sufficient fragments of moulded or worked masonry may be collected to enable a window or door surround to be reconstructed. It should, however, be pointed out that such a course of action needs to be supported at least by documentary evidence of the pre-existing assemblage.

EX-SITU FRAGMENTS

Fragments of stone, and occasionally of timber, may be found underground and in later buildings and walls, and in piles of clearance material around a site. Old building timbers may have been reused as lintels, pieces of wall frame, and roofing in later buildings.

Architectural and sculptural stones, including dressed cornerstones, window and door jambs, window tracery, arch stones, effigies, cross-inscribed slabs, crosses and bullauns may be found reused in stone walls and in scatters and piles of cleared stone.

Any such fragments found on or around the site should be brought to the attention of the conservation professional or archaeologist, and the location where they were found should be recorded. Occasionally, architectural details such as surviving plasterwork on walls may be discovered. Such finds should be left undisturbed until expert advice has been received.

Stone fragments taken from a ruin may be found nearby. Here, a piece of the mediaeval window tracery was taken from a ruined church, possibly centuries ago, and used as a grave marker. While the location of the stone fragment should be identified and recorded, it would not be appropriate to move it from its current location

A programme of repair and conservation was carried out at Kell, Lemanaghan by Offaly County Council, grant-aided by the Heritage Council and the Department of the Environment, Heritage and Local Government. While the precise date of the building is not known, it is almost certainly pre-Norman. Here the range of images shows the process of conservation (anti-clockwise from top left) – the initial condition and the heavy growth of ivy removed. Research produced the antiquarian drawing and the early photograph which informed the repair of the doorway. The final image (top right) shows the completed repair of the west gable. The repair work on site was carried out under the supervision of stonemason, Willie McErlean, seen in the bottom left photograph. He was part of the team which prepared this text and, sadly, died suddenly before publication

(From JRSAI Volume XIII, 1874)

(Image courtesy of Offaly Historical and Archaeological Society)

7. Other Important Issues

In addition to the actual process of identifying problems and carrying out repairs, there are other important issues to be considered in approaching the conservation of a ruined structure.

Setting

A ruined structure should not be considered in isolation from its setting. Any repairs, conservation, ancillary works or planting should be planned so as to respect the wider environment and ecology of the area. The following issues should be taken into consideration in any proposals:

> Wholesale clearance of vegetation around a site may lead to a loss of important flora and fauna and may be illegal if undertaken during the bird-nesting season or where it involves the disturbance of wildlife such as roosting bats

> Planting of trees and other vegetation for amenity purposes should be carefully considered so as not to create potential damage to the ruin in future years, disturb the indigenous flora and fauna, or introduce inappropriate species onto the site

> Ancillary buildings and fencing should be sited and designed so as not to create an adverse impact on the surrounding landscape

> Site services should not disturb underground archaeology in the area around the site

Floodlighting

Floodlighting is sometimes used to enhance significant landmark structures and to highlight particular architectural features on historic buildings. However, the benefits of any proposal to floodlight a structure should be balanced against national policies to conserve energy and reduce carbon emissions. In addition, consideration should be given to the effects of light pollution particularly in rural locations. Floodlighting schemes require ongoing maintenance including regular inspections, cleaning of fittings and replacement of lamps and faulty luminaires. This should be taken into consideration and planned for in any proposal. In all cases where floodlighting is proposed to a monument included in the RMP, the National Monuments Service should be consulted.

Where the structure proposed to be floodlit is a protected structure or is located in an architectural conservation area, planning permission is likely to be required. A proposed lighting scheme should be carefully planned to minimise the impact of cabling and light fittings fixed on or near the structure and to ensure that the floodlighting enhances rather than detracts from the appearance of the structure. Floodlighting installations have the potential to impact on archaeological heritage through additional electricity poles, underground cabling and light installations requiring excavation and on natural heritage by affecting the activity rhythms of both plants and animals. In principle, lighting should not be used on a structure where a bat colony is in occupation. If a structure is suspected of having a bat roost, a specialist survey of the structure will be required and advice sought from the National Parks and Wildlife Service of the Department of the Environment, Heritage and Local Government.

The number of lights should be kept to a minimum and lighting should never be left on all night. Dusk to midnight is a generally suggested timeframe, subject to the recommendations of the National Parks and Wildlife Service in a particular case. Cross-lighting and back-lighting should also be considered as an alternative to up-lighting to avoid light pollution of the night sky.

Lightning protection

Certain structures are at very high risk of being struck by lightning, depending on their location, height, or proximity to water. Indeed, many structures have become ruinous as a result of a lightning strike. There is an assessment procedure for the degree of risk from lightning and there are established codes setting out the standards. A system of lightning protection may be considered necessary following the assessment and may involve conductors at high level and copper downtapes to ground level. This area requires specialist expertise.

Ruins are often found in isolated areas and a structure such as a tower may be the tallest in its locality making it vulnerable to lightning strike. The tower of this ruined church was struck by lightning which destroyed the corner pinnacle

Ecological and wildlife issues

Some historic ruined sites may contain rare or protected plant species. These need to be examined and recorded before any disturbance takes place as there may be rare plants and plants of historical significance such as fruit trees and culinary or medicinal herbs surviving from a dwelling or monastic foundation.

Significant plants may be found growing on historic sites. This is a rare grass, ratstail fescus, which has been identified at several mediaeval sites

Vegetation can provide a valuable habitat and food source for wildlife, including protected species, such as bats, nesting birds, and the like which in turn may support the survival of barn owls and other such species. The protection of these habitats must be balanced against the damage being caused by the weight of the vegetation or by branches scraping masonry walls. If similar habitats are available nearby, there may be no loss or undue disturbance if the timing of the removal is well-planned. Defects in a structure such as missing stones or gaps in pointing may be entrances to roosts for bats or may contain the nests of small birds. Where there is evidence such as bat droppings below the hole, staining on stonework and the like, the first step should be to have a bat survey carried out by an appropriately-qualified bat expert. Where bats are present or there is evidence that they have used, or are using a ruin, the National Parks and Wildlife Service should be contacted for informed advice and guidance before any works are programmed and initiated. If there is an active bat roost, works will need to be programmed to cause the minimal amount of disturbance and measures put in place to allow bats to continue to use the site upon completion.

Burrowing animals such as rabbits and badgers can destabilise the ground around historic ruins and cause damage to underground archaeology. However, badgers are a protected species and any damage to a sett should be avoided and may be illegal.

It should be remembered that stone exposed at the earth's surface is not just an inanimate surface. It is a habitat which is colonised by lichens of many kinds, and the richness of this lichen community increases with time. For this reason the headstones in old graveyards can often harbour important lichen species, which may be obliterated by indiscriminate cleaning. Where blocks of stone are cemented by mortar, the wall thus formed is a microhabitat dominated initially by lichens and mosses, and later colonised by a multitude of small animals. Apart from its possible ecological value, this veneer of vegetation often imparts a considerable aesthetic character to a wall.

Farming practices

The presence of grazing animals in and around masonry ruins can have a significant damaging effect. The ruins are an attraction for animals seeking shelter from inclement weather. The presence of animals softens up and poaches the ground, resulting in water lodging, while their manure may accelerate damaging plant growth. Large animals can cause damage or deterioration by pushing through narrow openings or displacing and dislodging masonry thereby putting the animals at risk of injury, and by scratching against stones. In addition, fallen masonry which may have worked or sculpted surfaces can be destroyed by heavy animals passing back and forth. Fencing off structures from animals can be effective, but it will be necessary to introduce a ground maintenance regime within the fenced area.

The stones that formed the jambs to this doorway have become dislodged because of cattle pushing through the narrow opening

Structures accessible to the public

Some ruined structures are located in an area accessible to the public, such as graveyards. In these circumstances, it is advisable that the owner, usually the local authority, should prepare a management plan which will address issues such as ground surface preparation, public safety, interpretative information and signage. If the site is a recorded monument, early discussion with the Department of the Environment, Heritage and Local Government is highly advisable as agreement, and in some cases, Ministerial Consent will be required.

Burials

A local authority may have complex issues to resolve where burials are still taking place and causing damage to adjacent walls. It may be necessary to close a graveyard to additional burials, or forbid disturbance of any ground close to walls or in vulnerable areas.

Reuse of ruins

It is not the purpose of this guide to provide advice on the restoration of ruinous structures. The intention of this section is to provide a broad outline of the types of issues that will arise when such works are proposed.

While the aim for most structures is to preserve them in the state in which they have come down to us, in some cases restoration to active use may be the most viable way to ensure their continued existence. Such structures might include tower houses, mansions, cottages or churches. It is important that any proposed additions and alterations be firmly based on architectural, historical, structural and archaeological evidence. The impact of the works on the wider landscape and any ecological impact must be considered and mitigated. Restoration of a ruin for reuse should not be undertaken lightly. It is certain to be a lengthy and difficult project and will require expert advice from the earliest stages of the process. Early consultation with the relevant statutory authorities is advisable.

Among the issues that should be given consideration are the following:

> The importance of the ruin – some ruins are too important for reuse to be considered. This should be established using expert advice
> The scenic value of the ruin in the landscape may make it difficult to successfully extend or otherwise alter the external appearance of the structure
> The new use should be similar or close to the original use. The character and special interest of the structure should not be damaged
> Too many functions should not be fitted into a space
> Restoration works should be based on surviving architectural evidence avoiding conjecture. Careful research is needed in advance of historic descriptions, drawings, photographs and the like and the results of the research should inform the decisions taken
> A high-quality, architecturally-creative solution may sometimes be appropriate
> The provision of services can be difficult. The use of existing routes through the building such as flues may be an option
> Septic tanks and service pipes may have an impact on sub-surface archaeology
> Vehicular access and landscaping around the building may have an effect on the character and setting
> The design and location of any proposed ancillary buildings or extensions will require very careful consideration and in some cases these may not be appropriate
> The external walls of the building may need to be rendered to make the building habitable
> Historical walls and floors may not have the structural capacity for a new use
> Parapet walls of structures such as tower houses may require some rebuilding to facilitate re-roofing and this may not be considered appropriate in some cases

Ballycowan Castle, County Offaly, a fortified house of c.1589 which was added to and refurbished by Sir Jasper Herbert in 1626

Useful contacts

If the ruin is a protected structure, the architectural conservation officer in the local authority should be the first person to contact with queries regarding works to it. If the structure is included on the Record of Monuments and Places, the National Monuments Service of the Department of the Environment, Heritage and Local Government should be the first point of contact. Other useful contacts include:

Department of the Environment, Heritage and Local Government, Custom House, Dublin 1
> Architectural Heritage Advisory Unit
> National Monuments Service
> National Parks and Wildlife Service

Telephone: 01 888 2000
Web: www.environ.ie
www.buildingsofireland.ie
www.archaeology.ie
www.npws.ie

Building Limes Forum Ireland
Web: www.buildinglimesforumireland.com

Construction Industry Federation, Register of Heritage Contractors, Construction House, Canal Road, Dublin 6
Telephone: (01) 406 6000
Web: www.cif.ie and www.heritageregistration.ie

Engineers Ireland, 22 Clyde Road, Ballsbridge, Dublin 4
Telephone: 01 665 1300
Web: www.iei.ie

Heritage Council, Áras na hOidhreachta, Church Lane, Kilkenny, Co. Kilkenny
Telephone: (056) 777 0777
Web: www.heritagecouncil.ie

Irish Architectural Archive, 45 Merrion Square, Dublin 2
Telephone: (01) 663 3040
Web: www.iarc.ie

Royal Institute of the Architects of Ireland, 8 Merrion Square, Dublin 2
Telephone: (01) 676 1703
Web: www.riai.ie

Further reading

Allen, Geoffrey; Allen, Jim; Elton, Nick; Farey, Michael; Holmes, Stafford; Livesey, Paul and Radonjic, Mileva. *Hydraulic Lime Mortar for Stone, Brick and Block Masonry.* Shaftsbury: Donhead Publishing Ltd. (2003)

Ashurst, John and Nicola. *Practical Building Conservation, Volumes I, 2 & 3.* Hampshire: Gower Technical Press (1988)

Department of the Environment, Heritage and Local Government. *Architectural Heritage Protection – guidelines for planning authorities.* Dublin: Stationery Office (2004)

Department of the Environment, Heritage and Local Government. *National Policy on Town Defences.* Dublin: DoEHLG (2008) Available to download from www.environ.ie

Feilden, Bernard M. *Conservation of Historic Buildings.* Oxford: Architectural Press. First published 1982. Third edition (2003)

Historic Scotland. *Preparation and Use of Lime Mortars. Technical Advice Note 1.* Edinburgh: HMSO (1995)

Holmes, Stafford and Wingate, Michael. *Building with Lime: a practical introduction.* Rugby: Intermediate Technology Publications (1995, revised 2002)

International Council on Monuments and Sites (ICOMOS) *International Charter for the Conservation and Restoration of Monuments and Sites (Venice Charter).* Adopted at Venice (1966)

ICOMOS, *Charter for the Conservation of Places of Cultural Significance (Burra Charter).* Adopted at Burra, Australia (1979, revised 1999)

Lalor, Brian. *The Irish Round Tower, Origins and Architecture Explored.* Cork: The Collins Press (1999)

Leask, Harold G. *Irish Castles and Castellated Houses.* Dundalk: Dundalgan Press Ltd. (1941)

Leask, Harold G. *Irish Churches and Monastic Buildings* Volumes 1-3. Dundalk: Dundalgan Press Ltd (1955)

McAfee, Patrick. *Irish Stone Walls.* Dublin: O'Brien Press (1997)

McAfee, Patrick. *Lime Works.* Dublin: Building Limes Forum of Ireland and Associated Editions (2009)

McAfee, Patrick. *Stone Buildings.* Dublin: O'Brien Press (1998)

Manning, Conleth (ed.). *From Ringforts to Fortified Houses – studies on castles and other monuments in honour of David Sweetman.* Bray: Wordwell (2007)

National Monuments and Historic Properties Service, Office of Public Works. *The Care and Conservation of Graveyards.* Dublin (1995)

Pavía, Sara and Bolton, Jason. *Stone, Brick and Mortar.* Bray: Wordwell Ltd. (2000)

Glossary

AGGREGATE

Material such as sand or small stones used, when mixed with a binder and water, to form a mortar or concrete

ARMATURE

A concealed light reinforcement cage, generally for slender elements such as columns or tracery

ASHLAR

Cut stone worked to even faces and right-angled edges and laid in a regular pattern with fine joints

BED JOINT

The horizontal mortar joint between stone or brick courses

BLOCKING COURSE

The course of masonry erected above a cornice to anchor it both visually and structurally

CEMENT

A binding material mixed with aggregate and water to form a mortar or concrete. The term is usually taken to mean an artificial cement such as Ordinary Portland Cement (OPC)

CHEVAUX DE FRISE

Closely set, sharp, upright stones or stakes set at various angles forming a defensive structure around a fort

CONSERVATION

All the processes of looking after a place so as to retain its cultural significance (from the 'Burra Charter')

CONSOLIDATION

The addition of new material into the fabric of a building, to ensure its continued survival

COPING

A capping or covering to the top of a wall to prevent water entering the core of the wall

COURSE

A horizontal layer of stones or bricks together with its bed joint

CRAMP

A metal strap or pin built into a wall to hold together elements such as adjacent blocks of stone

DAMP-PROOF COURSE OR DPC

An impervious layer built into a wall a little above ground level to prevent rising damp. A DPC can also be used below window sills, above lintels and beneath coping stones to prevent water penetration of the interior of the building

DRESSINGS

Moulded masonry architectural features to a façade such as door and window architraves, string courses and quoins

FLASHING

A flat sheet of impervious material, usually lead, zinc or copper, covering the junction between materials or elements of a building to prevent water penetration

FLAUNCHING

A sloping mortar fillet, such as around the base of a chimney pot to hold it in place, or to a wall top to throw off rainwater

FOLDING WEDGES

Wedges are tapered pieces of timber used to secure joints. When used in pairs, with their tapered slopes opposing, they are called folding wedges and are used for levelling and easing temporary supports and formwork

FRENCH DRAIN

A trench filled with gravel or other loose material to collect ground water and deflect it away from a building

GABLE

The area of wall at the end of a pitched roof between the level of the eaves and the apex, usually triangular in shape

INDENTING

The process of replacing a damaged stone or part of a stone by inserting a piece of new matching stone

INTRADOS
The interior curve of an arch

JOINT
The mortar between two stones or bricks

LIME, HYDRAULIC
Hydraulic limes contain a percentage of clay which produces a pozzolanic effect in mortars, that is, the mortars set chemically assisted by the presence of water. Hydraulic limes can occur naturally, or can be artificially made

LIME MORTAR
A mortar made from lime, aggregate and water that carbonises and hardens on exposure to air

LIME, NON-HYDRAULIC
Non-hydraulic limes are pure, or almost pure, lime. Mortars made of non-hydraulic limes can only set through contact with air, a process known as carbonation

LIME PUTTY
A soft putty made from slaking quicklime in water. Used as a binder in most traditional mortars and renders prior to the invention of Portland cement

LIMEWASH
A form of thin lime putty used as a paint or protective coating. It differs from whitewash which is a mixture of chalk and water that does not carbonate

LINTEL
A small beam made of wood, stone or concrete which spans the top of an opening such as a door, window or fireplace and supports the wall above

MAINTENANCE
The continuous protective care of the fabric and setting of a place, and is to be distinguished from repair. Repair involves restoration or reconstruction (from the 'Burra Charter')

MASON'S MARK
A symbol or initial cut into stonework by the mason executing the work. Usually associated with mediaeval masonry

MITRE
Join at the corner of a moulding, usually at a 45 degree angle, in order that the individual pieces may fit together

MORTAR
The mixture of a binder (such as lime or cement), aggregate and water to form a substance used to bind stones or bricks together in a masonry wall

PARAPET
The part of a wall that rises above a roof or terrace

PINNINGS
Also known as 'gallets' or 'spalls'. Small pieces of stone or other material pressed into the mortar joints of a wall either as decoration or to reduce the amount of mortar required and thus reduce the danger of shrinkage

PLINTH
The projecting base of a wall or column

POINTING
This term is used in two ways – to describe the application of facing mortar onto the bedding mortar, and to describe the actual material used

PORTLAND CEMENT
Artificial cement invented by Joseph Aspdin in 1824 and so called because of its perceived resemblance to Portland stone. It sets rapidly and is very hard when set

POZZOLAN
A type of naturally-occurring volcanic ash, or any artificial substitute for it, added to a mortar to achieve a quick, strong hydraulic set

PRESERVATION
Maintaining the fabric of a place in its existing state and retarding deterioration (from the 'Burra Charter')

QUOIN
A dressed stone forming the corner of a building, often decorated or raised

RANDOM RUBBLE
Stones of irregular size and shape used for building purposes

RECONSTRUCTION
Returning a place to a known earlier state and is distinguished from restoration by the introduction of new material into the fabric (from the 'Burra Charter')

RELIEVING ARCH
An arch built into the masonry above an opening to reduce the masonry's downward force upon the lintel

RENDER
A mixture of a binder (such as lime or cement), an aggregate and water to form a coarse plaster which is applied to the external surfaces of walls

RE-POINTING
The replacement of mortar in the face joints of brickwork following either the erosion of the original mortar or its removal through raking out

RESTORATION
Returning the existing fabric of a place to a known earlier state by removing accretions or by reassembling existing components without the introduction of new material (from the 'Burra Charter')

REVEALS
The sides of an opening for a door or window, between the frame and the face of the wall. If cut at an angle, it may be called a splayed reveal

REVERSIBILITY
The principle of carrying out works to a building in such a way that they may be undone at a future date without inflicting damage to the fabric of that building

RIBBON OR STRAP POINTING
Pointing which is not flush with the building surface but stands proud

ROMAN CEMENT
Cement derived from a hydraulic clay. Used as a substitute for stone carving and as a render

ROUGH RACKING
The finishing of a wall end or wall head where the stones have exposed edges, not laid to line, but the mortar joints are well-finished to protect the core

RUBBLE WALL
Stone wall built with undressed masonry

RUSTICATION
In Classical architecture, the treatment of a wall surface, or part thereof, with strong texture to give emphasis and/or an impression of strength

SNECKED MASONRY
Sometimes called random ashlar, snecked masonry is built in discontinuous courses with small stones, or snecks, introduced to break courses

SOFT CAPPING
A method for protecting wall-tops by applying an earth layer such as sods over a mesh to seal the masonry

SPALLING
The gradual breaking away of small chips or flakes from the surface of individual bricks or stones

STRING COURSE
Decorative horizontal band of moulding found on an external wall, often at first floor level

TRACERY
Ornamental intersecting timber or stone mullions and transoms in a window, panel or vault. Typical of buildings built in the Gothic or Gothic-Revival styles

VERMICULATION
A form of rustication where the surface of stone is finished with shallow, winding channels to appear as if eaten by worms

VOUSSOIR
A wedge-shaped stone or brick forming part of an arch. The middle voussoir is called a keystone and is often carved and decorated